Also by Heinz R. Pagels

The Cosmic Code: Quantum Physics as the Language of Nature

*The Dreams of Reason: The Computer and the Rise of
the Sciences of Complexity*

Perfect Symmetry

The Search for the Beginning of Time

Heinz R. Pagels

Simon & Schuster Paperbacks
New York • London • Toronto • Sydney

Simon & Schuster Paperbacks
A Division of Simon & Schuster, Inc.
1230 Avenue of the Americas
New York, NY 10020

This Simon & Schuster trade paperback edition May 2009

SIMON & SCHUSTER PAPERBACKS and colophon are registered
trademarks of Simon & Schuster, Inc.

For information about special discounts for bulk purchases,
please contact Simon & Schuster Special Sales at
1-866-506-1949 or business@simonandschuster.com.

The Simon & Schuster Speakers Bureau can bring authors
to your live event. For more information or to book an event,
contact the Simon & Schuster Speakers Bureau at 1-866-248-3049
or visit our website at www.simonspeakers.com.

Manufactured in the United States of America

10   9   8   7   6   5   4   3   2   1

The Library of Congress has cataloged the hardcover edition as follows:
Pagels, Heinz, R., 1939–1988
     Perfect symmetry: the search for the beginning of time; Heinz R.
Pagels.
          p.   cm.
     Includes bibliographical references.
     Includes index.
     1. Cosmology.   2. Astrophysics.   3. Field theory (Physics)
   4. Space and time.   I. Title.
QB981.P25   1990
523.1—dc20   90-683
                                                              CIP

ISBN-13: 978-1-4391-4888-4
ISBN-10:   1-4391-4888-0

FOR ELAINE

Acknowledgments

In preparing this book I have been fortunate in having friends and colleagues who can offer open criticism or who have made suggestions that found their way into the text. I have benefited from comments by Jeremy Bernstein, John Brockman, Malcolm Diamond, John Faulkner, Randall Furlong, George Greenstein, Alan Guth, Edward Harrison, Joseph H. Hazen, Nicolas Herbert, James McCarthy, Richard Ogust, Jim Peebles, Anthony Tyler and Anthony Zee. I am especially grateful for the detailed criticism of George Field and Engelbert Schucking in the sections of the book dealing with astrophysics and cosmology. Alice Mayhew and Catherine Shaw did the major editorial work on the text and helped turn my English into English. Matthew Zimet's inventive illustrations delight the eye and do much to enhance the text. Finally, I want to thank the Board of Governors of The New York Academy of Sciences for their sympathetic appreciation of my interest in science writing.

Contents

• CONTENTS

In the beginning God created the heavens and the earth. The earth was without form and void and darkness was upon the face of the deep; and the Spirit of God was moving over the face of the waters.

—Genesis

Foreword

The children's books that were my first contact with the universe opened my imagination to thoughts of voyages to the moon, planets and stars. When I was older, however, I visited the Fels Planetarium in Philadelphia and the Hayden Planetarium in New York, and that simple, self-centered perception was shattered. The drama and power of the dynamic universe overwhelmed me. I learned that single galaxies contain more stars than all the human beings who have ever lived, and I saw projections of clusters of such galaxies moving in the void of space like schools of fish swimming in the sea. The reality of the immensity and duration of the universe caused a kind of "existential shock" that shook the foundations of my being. Everything I had experienced or known seemed insignificant placed in that vast ocean of existence.

While my sense of awe at the size and splendor of the universe is a feeling that has never quite left me, reflecting back on my childhood experience I see that the universe provided a screen upon which I could project my feelings about the immensity of existence; that external ocean mirrored the one within me. Later, as I pursued the study of theoretical physics at Princeton and Stanford Universities, my attitude toward the universe altered. The universe became less a screen for the projection of my feelings and more a puzzle challenging me as a scientist, a puzzle which left scattered, complex clues to its solution. The universe, in spite of its size, is a physical entity governed by the laws of space, time and matter. Someday (and that day is not yet here) physicists may know the laws that describe the creation of the universe and its subsequent evolution. The logical account of the foundations of physical existence will then be complete.

As we embark on the study of the universe, it is worth reminding ourselves that not so long ago, at the beginning of this century, physicists were puzzled by the properties of atoms. Atoms were so small (a few eminent scientists even doubted their existence) and behaved in such sporadic, uncontrollable ways that some people thought they lay beyond the power of scientific comprehension. Yet after major experimental and theoretical discoveries, physicists in the 1920s invented the quantum theory which explicated the weird world of the atom. New and unfamiliar physical concepts were incorporated into the quantum theory, concepts that have survived to the present day.

Similarly, as physicists attempt to comprehend the origin and evolution of the universe, they will certainly need to invent new and unfamiliar concepts. Scientists do not yet understand the fundamental laws that describe the very origin of the universe, at least not as well as they understand the laws describing atoms. But many scientists today are excited because such an understanding is currently in the making, a result of the intellectual synthesis of two scientific disciplines: quantum theory, which specifies the laws of the smallest things—the quantum particles—and cosmology, which specifies the laws that govern the largest thing—the entire universe.

A major reason for the growing intimacy between quantum physics and cosmology is the success of the "big bang" theory of the early universe. According to this theory, if we imagine going backward in time then we would see the universe contract, the galaxies move closer together until they meld into a hot, uniform gas of all the constituents of matter—the quantum particles—interacting at enormous energy. Elucidating the properties of such a gas of hot, interacting quantum particles is the purview of modern quantum theory. Physicists estimate that the high temperatures and high energies among the quantum particles eventually encountered in the early universe are physically unique—they become so high that they cannot be reproduced in laboratories here on earth. Hence the only possible "laboratory" that can test theories of quantum-particle interactions at ultrahigh energies is the universe itself.

Another reason for the growing intimacy between quantum theory and astronomy is that astronomers are now observing exotic objects like neutron stars, consisting of matter compressed to enormous densities, and possibly black holes, in which the very fabric of space and time undergoes unusual distortions. Like the early universe, these strange objects present extreme physical conditions that cannot be reproduced here on earth. Since it is the properties of space, time and matter, especially under extreme conditions, that physicists endeavor to understand, these new objects provide yet additional extraterrestrial laboratories for testing physical laws.

Were I to summarize the optimistic theme of this book in a single sentence, that sentence would be "From microcosm to macrocosm, from its origin to its end, the universe is described by physical laws comprehensible to the human mind."

I believe that physicists will someday soon understand the basic laws of the quantum creation of the universe (most probably out of nothing whatsoever) as well as astrophysicists now understand the interiors of stars. The universe, whose very mention invokes a sense of transcendence, will be comprehended as subject to natural laws like all other material things. In spite of its immensity and age, the universe will never seem the same.

Such a fulfillment of the program of the natural sciences will have a profound impact on human thinking. As knowledge of our universe matures, that ancient awestruck feeling of wonder at its size and duration seems inappropriate, a sensibility left over from an earlier age. Thousands of years ago, many people perceived the sun as a divine presence; today many people perceive the universe as essentially beyond human comprehension. But just as the sun is now understood in terms of astrophysical processes, so too will the universe be similarly understood. In the past, myths and the religious creation stories shaped the values of people who believed in them; likewise the emergent scientific cosmology will shape the values of those who accept it. Through the agency of scientific discovery the external order of the universe influences our consciousness and values.

This book is divided into four parts. The first part, "Herschel's Garden," gives the reader an overview of the dynamic universe discovered by astronomers—the stars, white dwarfs, neutron stars, black holes, interstellar gas and dust, quasars, galaxies, their distribution in space as clusters and superclusters of galaxies, and the cosmos as a whole. From this part of the book the reader should derive a sense not only of the size of the universe and our knowledge of its inhabitants but also of the puzzles confronting modern astronomy such as how stars are born and galaxies are evolving. I discuss some suggested solutions to these and other astronomical puzzles to which we can hope to achieve a final resolution as new observational data are acquired. Today, the search into the universe continues with instruments like satellites and radio telescopes, a search manifesting, in the words of the American astronomer Edwin Hubble, an "urge . . . older than history."

While the first part of the book describes the universe observed in space, the following two parts of the book describe a conceptual exploration of the universe in time. The second part, "The Early Universe," describes the remarkable picture of the universe when it was only seconds and minutes old—the "hot big bang," a theory that came about by the application of the laws of quantum-particle physics to the entire universe. Without using complicated mathematics I describe the basic framework for thinking about the quantum particles—the discipline known as "relativistic quantum-field theory"—and how it applies to the study of the early universe. Amazingly, physicists understand the universe better when it was seconds and minutes old than for either earlier or later times because when it was seconds old the universe was a uniform, rather simple, gas of quantum particles, whose properties are known. The early universe is better understood than the weather is today.

But the very success of the hot-big-bang theory gives physicists the confidence to press onward and conceptually explore the universe *before* the first nanosecond (one-billionth of a second) to the very origin of the universe. The third part of the book, "Wild Ideas," leaves the secure territory explored by astronomical observation and by high-energy laboratory experiments and speculates about the nature of that universe before the first nanosecond. I discuss "wild ideas" in the conceptual repertoire of theoretical physicists that might explicate the dynamics of the very early universe, ideas such as GUTs —grand unified theories—magnetic monopoles, supersymmetry and the world of many extra dimensions. If these ideas are correct—and many physicists think they are—then an amazing picture of the very early universe results.

The universe begins in a very hot state of utmost simplicity and symmetry and as it expands and cools its perfect symmetry is broken, giving rise to the complexity we see today. Our universe today is the frozen, asymmetric remnant of its earliest hot state, much as complex crystals of water are frozen out of a uniform gas of water vapor. I describe the inflationary universe—a conjectured pre–big-bang epoch of the universe, which may explain some puzzling features of the contemporary universe, such as its uniformity and age, as well as provide an explanation for the origin of the galaxies. The penultimate chapter of this third part of the book—as far as speculation is concerned—describes some recent mathematical models for the very origin of the universe—how the fabric of space, time and matter can be created out of absolutely nothing. What could have more perfect symmetry than absolute nothingness? For the first time in history, scientists have constructed mathematical models that account for the very creation of the universe out of nothing.

There is a short fourth part, "Reflections," which expresses my opinions and attitudes (not that the other parts of the book do not contain many of my opinions or intellectual biases as a theoretical physicist). Here the reader will find a chapter developing the metaphor of the universe as a cosmic computer for which the quantum particles are the "hardware," the laws of physics the "software" and the evolution of the universe is the execution of the program. In a final chapter called "First-Person Science," I explore the thoughts and feelings that a few people have had about the meaning of our strangely coherent universe.

New York, New York
Felton, California
1984

One

●

Herschel's Garden

The most beautiful and deepest experience a man can have is the sense of the mysterious. It is the underlying principle of religion as well as of all serious endeavour in art and in science. . . . He who never had this experience seems to me, if not dead, then at least blind. The sense that behind anything that can be experienced there is a something that our mind cannot grasp and whose beauty and sublimity reaches us only indirectly and as feeble reflexion, this is religiousness. In this sense I am religious. To me it suffices to wonder at these secrets and to attempt humbly to grasp with my mind a mere image of the lofty structure of all that there is.

—Albert Einstein, 1932

1

Herschel's Garden

There are two kinds of happiness or contentment for which we mortals are adapted; the first we experience in thinking and the other in feeling. The first is the purest and most unmixed. Let a man once know what sort of a being he is; how great the being which brought him into existence, how utterly transitory is everything in the material world, and let him realize this without passion in a quiet philosophical temper, and I maintain that then he is happy; as happy indeed as it is possible for him to be.
—William Herschel, from a letter to his brother Jacob

William Herschel, the greatest astronomer of the eighteenth century, began his career as a teenage oboist in the Hanoverian Foot Guards in a part of Germany then under the dominion of George II of England. Born in 1738, he wanted to become a professional musician and composer. However, at about the time of the battle of Astenbeck he was "so near to the field of action as to be within reach of gunshot." His father advised him to flee. To avoid the draft into regular military service, he left at age nineteen with his brother Jacob for England, where he pursued his career in music. In 1766, he was appointed organist at the Octagon Chapel in the resort town of Bath, where he also played in the Pump Room orchestra.

Not until he was thirty-five did Herschel's interest in astronomy begin. In Bath he bought many books on astronomy. Aided by his sister, Caroline, and brother Alexander, he made a fine reflecting telescope using a foundry he built in his house. No doubt his skill with musical instruments served him well in the construction of the precision instrument. Training this telescope at the sky, he discovered a new planet—Uranus—which he at first thought was a comet. Since ancient times the only known planets had been the six observable by the unaided eye. No one had anticipated an additional planet, and the shock of this discovery made Herschel and his telescope instantly famous. Not at a loss to express his gratitude to his adopted country, he called the new planet *Georgium Sidus* (George's Star) in honor of King George III, but later the name was changed. Herschel was elected to the Royal Society of London, George III became his patron and his career in astronomy was launched.

Herschel's entry into astronomy was not unusual—many great observational astronomers began their careers in different professions with only an ancillary interest in astronomy. But after making the major astronomical discovery of a strange new planet, Herschel found it difficult to resist the urge to continue exploring the universe. The passion for science and the passion for music are driven by the same desire: to realize beauty in one's vision of the world.

Herschel's accomplishments in astronomy are all the more remarkable in retrospect; indeed, many of his observations and insights could not be fully appreciated until the twentieth century. He realized, for example, that because of the finite velocity of light we see distant celestial objects as they were in the past. As we look into the depths of the universe, we look at the way it was millions and billions of years ago when the light we are now receiving was first emitted. Remarkably, the universe contains the record of its past the way that sedimentary layers of rock contain the geological record of the earth's past. And that fact opened the window to the evolutionary view of the universe held today.

Herschel became obsessed with the problems of determining the structure of the Milky Way and locating the position of our sun within it. He was even convinced that some nebulae were both external to the Milky Way and similar to it—thus anticipating the "island universe" theory of galaxies. Because it was impossible to estimate distances to the stars with the techniques available to Herschel, his picture of the Milky Way galaxy was quantitatively wrong. Much to his credit as a scientist, but to his personal disappointment, he later abandoned his picture of the Milky Way as a large disk (which is in fact correct) when he realized that his observational methods were inadequate for the task of accurately establishing its shape. But he was the first to show that the Milky Way stars are not symmetrically arranged about the sun—an important fact substantiated by modern observations. He thus destroyed forever the idea of the heavens as a celestial sphere surrounding the sun. In spite of subsequent speculations, no further progress on this problem was made until Harlow Shapley, the American astronomer, published his studies on the shape of the Milky Way some 140 years later.

On the day of his election to the Royal Society, Herschel was sent a copy of the new catalogue of 103 nebulae published by Charles Messier and Pierre Méchain by his friend Dr. W. Watson, Jr. He immediately began to train his wonderful telescope upon these strange objects, hoping to discover a few more that might have been missed. Instead, he discovered two thousand new nebulae and began a list of his own. This was the beginning of a new catalogue (to which his son, John, added the many more nebulae he observed in the southern hemisphere some years later) and formed the foundation of all the modern catalogues of galaxies.

Herschel also discovered many double-star systems—two stars in orbit about each other—and showed that they obey Newton's law of gravitation. Today we know that about half of all observable stars are members of such binary systems. Herschel's discovery that Newton's law applies to the movement of faraway stars, and not just to the movement of planets about the sun, was pivotal. He also showed that the sun, rather than being fixed in space, actually moves, in this case toward the star Lambda Herculis—a revolutionary idea comparable to Copernicus' declaration that the earth moves about the sun. Like many of his contemporaries, Herschel thought the moon, the planets and the sun were inhabited (he thought there was a cool surface under the sun's hot atmosphere). Perhaps no person before or since has spent so much time looking through a telescope.

Herschel made a great conceptual shift in astronomy. Previously people shared a Newtonian, mechanical view of the stars as subject only to the force of gravity. But Herschel, thoroughly in tune with our modern view, suggested that other dynamic processes were shaping the universe. In the baroque style of his time, he writes about the possibility of old stars colliding to form new ones:

> If it were not perhaps too hazardous to pursue a former surmise of a renewal in what I figuratively call the Laboratories of the Universe, the stars forming these extraordinary nebulae, by some decay or waste

of nature, being no longer fit for their former purposes, and having their projectile forces, if any such they had, retarded in each other's atmosphere, may rush at last together, and either in succession, or by one general tremendous shock, unite into a new body. Perhaps the extraordinary and sudden blaze of a new star in Cassiopeia's chair, in 1572, might possibly be of such a nature.

Herschel appreciated the vast variety of the heavens—even in his time, when the observed universe was far simpler than what we behold today. He saw the universe as a changing, evolving place and said that examining the stars was like examining a large garden in which some plants are old, others young, some are being born, others are dying. Although we may not see an individual plant growing, we do see lots of examples of that plant in all stages of its life, and that observation gives us a clue to understanding its growth. Likewise, the astronomer sees an evolutionary continuum in the development of stars, and perhaps in galaxies and clusters of galaxies, and that is his clue to the dynamics of change in the universe. Herschel wrote:

This method of viewing the heavens seems to throw them into a new kind of light. They are now seen to resemble a luxuriant garden, which contains the greatest variety of productions, in different flourishing beds; and one advantage we may at least reap from it is, that we can, as it were extend the range of our experience to an immense duration. For, to continue the simile I have borrowed from the vegetable kingdom, is it not almost the same thing, whether we live successively to witness the germination, blooming, foliage, fecundity, fading, withering and corruption of a plant, or whether a vast number of specimens, selected from every stage through which the plant passes in the course of its existence, be brought at once to our view?

The dynamic universe is Herschel's garden. We might press his analogy further. Botanists once studied plants only as isolated organisms. But as the life of plants became better understood, botanists realized that far from existing independently, each plant depends upon an ecological network, a complex environment, for its life.

Likewise with planets, stars and galaxies. While astronomers can study them independently, it is becoming clear that there is a complex interplay between and among all the objects we observe in the heavens. For example, the atoms of planets and the atoms in our bodies consist of many heavy chemical elements that were cooked up out of lighter elements in the nuclear furnaces of stars long ago. The rate at which new stars are born in the arms of a spiral galaxy influences the dynamics of the whole galaxy, which in turn influences star formation. Like life in a garden, life in the universe depends on a complex relation of parts to the whole. To see this relation, let us wander into Herschel's "luxuriant garden" and get a quick overview of what is there.

The heavens are alive with a great variety of celestial objects. Besides billions of stars similar to our sun, astronomers have discovered lots of very different kinds of stars. Among these is Betelgeuse, the "red supergiant" star in the constellation of Orion, a star so swollen it occupies a space as large as the earth's orbit. Because Betelgeuse is so large and so relatively near, it is the first star to have its disk resolved—we can actually see it as a circular disk, not a point of light, by using an optical viewing technique called speckle interferometry.

Astronomers have also discovered stars at the very end of their lives, white dwarfs and neutron stars. Eventually the sun will turn into a red giant and then, in turn, into a white dwarf, a tiny star shining with the last reserves of its energy. Stars more massive than the sun eventually undergo a more dramatic fate. Some such stars explode in a "supernova," releasing in a single second the equivalent of all the energy our sun will have released in its entire lifetime of billions of years. The "new star in Cassiopeia's chair, in 1572," to which Herschel referred was the first observation of a supernova explosion in the West. The remnant of this explosion is a tiny neutron star consisting of matter compacted down to the density of an atomic nucleus—several tons per cubic centimeter. Lots of refuse matter from this explosion is spilled out into space, contributing heavy elements to the interstellar gas. This matter eventually finds its way into making new stars in a gigantic recycling process. Although no one has actually seen a new star being made, astronomers know that birthplaces of stars are the dense gaseous nebulae such as the Orion nebula in the arms of our spiral galaxy.

Imagine flying out of the solar system, beyond the Milky Way, and looking back. What would we see? First, we would behold within the disk shape the great beautiful spiral arms of our galaxy, which contains new stars (like our sun) and lots of interstellar gas and dust. Farther away, we would see the arms twisting around and embracing a "central bulge," roughly spherical in shape, made of older stars and harboring in its core sources of immense energy—perhaps a gigantic black hole. And finally, looking above and below the plane of the disk, we would see the galaxy's "halo," at least as large as the disk and roughly spherical in shape, consisting of about a hundred sparsely distributed "globular clusters" of old stars gravitationally bound to each other and in orbit about the galaxy itself. What we could see of our galaxy would be only part of the story. There are other invisible components as well—infrared radiation, X rays, magnetic fields and subatomic particles. We now know that the galaxy is surrounded by a corona of hot gas and that most of the mass of a galaxy may be in the form of dark matter, not the visible stars and gas. Our galaxy is a complex dynamic entity we are only beginning to understand.

If we look at our galaxy from a yet wider and farther perspective, we see that it is adorned with smaller satellite galaxies—the seven "dwarf galaxies" and Leo I and II, other small galaxies—in orbit about it. In addition to these dwarf galaxies, sparse in stars and roughly spherical in shape, we would see lying close to our galaxy the Large and Small Magellanic Clouds, which are small, irregularly shaped galaxies. The Large Magellanic Cloud is being torn apart by gravitational tidal interactions with our galaxy. The evidence for this is the existence of the Magellanic Stream—a giant stream of gas connecting our galaxy with the Magellanic Cloud.

Taking in a still larger volume, we would see our neighbor the Andromeda Galaxy, another spiral similar to the Milky Way, with its own group of smaller satellite galaxies in orbit about it. There are other galaxies in our local group, all of them in the suburbs of a disk-shaped cluster of galaxies—the Virgo "supergalaxy." The Virgo cluster is but one of many such groups of galaxies. Clusters of galaxies tend to group into "superclusters" of galaxies. The visible universe contains at least 100 billion galaxies—a number beyond our everyday comprehension.

Nature has been generous to astronomers, offering an abundance of different stars and galaxies at all stages of their lives to look at. Because of that abundance, astronomers can put together a picture of a dynamic universe, plotting the lives of stars and the evolution of galaxies, even though no changes can be detected over a human life span.

Although nature has been generous in offering a variety of stars and galaxies, it has been even more generous in the allotment of space. Even astronomers are amazed at the size of the universe once they pause to reflect upon the meaning of the distances they are calculating. In spite of their vast numbers, stars do not begin to crowd each other because of the vastness of the space around them. If the sun were shrunk to the size of a pea, its nearest neighbor, Proxima Centauri, the binary partner of Alpha Centauri, would be about 90 miles away, and its next-nearest neighbor, Barnard's star, would be about 125 miles distant. That leaves lots of elbow room for stars. By contrast, if our entire Milky Way galaxy were shrunk to the size of a pea, its own nearest neighbor, the Andromeda galaxy, would be only 4 inches away. This is still lots of room—but galaxies do collide from time to time, especially in the dense clusters of galaxies like the Coma cluster where they are more crowded together.

Herschel's garden—the universe—is far larger than he could have imagined. Exotic new celestial objects recently discovered by astronomers would have excited him as they now excite us. The universe turns out to be far more peculiar than anyone could have imagined. Herschel's scientific progeny, extending consciousness to the ends of space and time, have created a new vision of reality.

Today scientists confront the universe as a puzzle with scattered clues to its solution. Challenging as it is, many believe that they will solve it someday. That day may be closer than many people think.

The first part of this book surveys the territory explored by astronomers, leading to the discovery of the modern universe itself. I organize this survey chapter by chapter, first exploring the stars, then moving on to galaxies, clusters and superclusters of galaxies and finally, to the immensity of the universe as a whole. As astronomers take in greater distance scales, they are also looking further into the past—a progression deeper into the universe roughly in step with the development of increasingly powerful instruments for astronomical exploration. I emphasize the most recent astronomical discoveries, but these revelations will be placed in the historical context of the great steps that went before.

Let us now look more closely at the objects in Herschel's garden— the stars, gas and galaxies—in order to know them as a good gardener knows his plants. Let us have a good look, for this beautiful garden is evolving, never again to be the same.

2

The Birth and Life of Stars

A scientist commonly professes to base his beliefs on observations, not theories. . . . I have never come across anyone who carries this profession into practice. . . . Observation is not sufficient . . . theory has an important share in determining belief.
    —Arthur S. Eddington, *The Expanding Universe,* 1933

Stars are born, they live and they die. Filling the night sky like beacons in an ocean of darkness, they have guided our thoughts over the millennia to the secure harbor of reason. It was in the attempt to understand the motion of stars and planets that the human mind first grasped the idea of natural law. But the stars are more than objects for scientific investigation. Like the sun and the moon, they are embedded in our unconsciousness—we sense their presence even if we do not see them.

Arrayed in an apparently random pattern in the sky, the stars provide a perfect screen for the projection of our feelings. In that pattern, ancient priests and poets saw the figures of myth and nature; the stars were gods—archetypes of permanence in an impermanent world. Compared with human life or the life of nations and empires, stars appear to live forever, indifferent to the passions of our existence. Yet somehow we feel that in spite of the immense distances which separate us from all stars save our sun, the destiny of humanity is profoundly intertwined with them. We hope that life on earth may share in the permanence of the stars, the galaxies and the universe itself. Whether that hoped-for permanence is no more than a projection upon the heavens of our modern myth of progress and therefore, like the ancient projections of the figures of myth, also an illusion, time will tell. The stars, like the gods they once represented, continue to play with our deepest feelings. But what are stars?

In this chapter, we will be taking an overview of what astrophysicists have learned about the birth and life of stars, with special emphasis on the most recent findings. The following chapter is devoted to describing the spectacular death of stars. Although we examine stars as if they were individuals, it is important to bear in mind that they are members of a larger society—the galaxy—which nurtures them in their birth, is the province of their life and receives their remains upon death. Stars may not exist outside of galaxies; such a lethal separation of the part from the whole would violate some principle of cosmic togetherness.

For a long time, people puzzling over the stars tried to understand how they gave off their light in terms of familiar physical processes, such as a burning fire. Centuries ago, Nicolas of Cusa and other philosophers speculated that the stars were but distant suns. If other stars were comparable to the sun, then the light and heat radiated by all the stars were very great indeed. What could fuel such immense radiations? No processes ever seen on earth could explain how stars burned. Not until Einstein, in the first decade of this century, showed that matter and energy are interconvertible and experimental physicists explored the atomic nucleus was an explanation possible.

Physicists, inspired by these new discoveries, suggested that the spontaneous emission of quantum particles from the atomic nucleus known as radioactivity represented such a transformation of matter into energy, but no detailed explanation of that process existed. Some of these ideas were taken up by Arthur S. Eddington, the English astronomer, in his influential book *The Internal Constitution of Stars*, published in 1926. Here, with great effect, he applied the newly discovered laws of atomic physics to the interiors of stars and outlined the central problems confronting astrophysicists, the scientists who study the physics of stars. The main problem was to find the source of stellar energy. Eddington boldly insisted that only subatomic nuclear processes could do the job. In his book he wrote, "The measurement of liberation of subatomic energy is one of the commonest astronomical observations; and unless the arguments of this book are entirely fallacious we have fair knowledge of the conditions of density and temperature of the matter which is liberating it."

According to Eddington's calculations, the center of a star like the sun had a temperature of 40 million Kelvin—very hot indeed. (More recent calculations indicate a temperature closer to 14 million Kelvin.) Since temperature in ordinary stars measures the energy of motion of microscopic particles, we would conclude that in the center of a star, atomic nuclei like that of the hydrogen nucleus, a single proton, would be very energetic; they would be moving extremely rapidly, and smashing into one another all the time. But the physicists of 1926 believed that if atomic nuclei got close to one another, they would repel each other. They would not fuse to form a heavier nucleus and liberate the needed nuclear energy. Even at the high temperature at the center of a star, the repulsive barrier preventing the contact was too high to surmount. Yet Eddington continued to insist that nuclear processes were responsible for a star's energy.

The breakthrough came in 1928 with the invention of the new quantum theory and the discovery by George Gamow, R. W. Gurney and E. U. Condon of what it implied—that particles did not have to surmount the repulsive energy barrier but could tunnel right under it. The energy required of nuclear particles to tunnel under the repulsive barrier was far less than that needed to surmount it. Now Eddington's guess could be made to work. In 1929, the physicists Robert d'Atkinson and Fritz Houtermans showed how this "tunneling effect" could explain the energy production of stars by nuclear fusion.

Yet, the precise nuclear reactions that might occur in the core of a star (consisting mostly of hydrogen and helium nuclei flying about) remained unknown. How, in detail, could the hydrogen nuclei fuse to eventually form the heavier nucleus of helium—a process called nuclear burning? In 1938, Hans Bethe in the United States and, independently, Carl Friedrich von Weizsäcker in Germany mathematically deduced the first of two nuclear reactions—the "carbon cycle" —which answered this question. They showed how, beginning with just hydrogen nuclei and a carbon nucleus as a catalytic agent, one could burn the hydrogen into helium, liberating immense energy. Bethe and Charles Critchfield demonstrated yet another way that hydrogen could burn into helium without the necessity of a carbon catalyst—the "proton-proton chain," also suggested by von Weizsäcker—and Bethe went on to prove that this reaction and the carbon cycle were the only possible ones. But, in spite of the discovery of the mechanisms of nuclear burning for the energy release in stars, scientists found it was a long and arduous task to show how this process in fact accounts for the observed properties of stars.

In the 1950s, computer modeling of stars began in earnest, and this provided a new method that would reveal the complex astrophysical consequences of elementary physical laws. The knowledge gleaned from the study of nuclear explosions at the Los Alamos Scientific Laboratory helped astrophysicists struggling to understand nuclear processes inside stars. They could now program on a computer the equations that described the interior of stars—the temperature and pressure, the complex nuclear reactions. Astrophysicists made great progress. In 1955, Fred Hoyle and Martin Schwarzschild made a breakthrough by using computer simulations of the evolution of an ordinary star to show how it turned into a bloated red giant star.

By and large, mathematical computer modeling of stars has been remarkably successful; today we have the makings of a theory of stars in good agreement with astronomical observation. Astrophysicists understand the major aspects of the evolution of stars from birth to death. The theory is far from complete; there are gaps, problems, observational puzzles. Yet the major successes are ground for optimism that a reasonably complete theory of stars and stellar evolution may be completed within this century. Some may even say it is already at hand. Stars are very complex entities, and scientists are still discovering new features of them—but they are small details compared with the major features already known and understood. Still, one cannot be satisfied until even the most bizarre behavior of stars is unraveled, and that may take more time.

During the last few decades, astrophysicists have discovered more about stars than was known in previous centuries, discoveries that were prompted by several major scientific advances, some of which have already been alluded to. First, the technology involved in astronomical observation made great strides. Sensitive electronic detectors that can "see" very faint objects; the advent of artificial satellites; the birth of X-ray astronomy and new optical, infrared and radio telescopes and their associated electronic systems have vastly improved observational capabilities. Second, the emergence of the quantum theory of atoms, the theoretical and experimental understanding of nuclear physics and plasma physics—the study of electrically neutral gases of charged particles—provided the theoretical foundations for modern astrophysics. Confident that they understood the laws of the microcosmic world of atomic and subatomic particles, scientists went on to build mathematical models of macrocosmic objects like stars. Third, high-speed computers enabled astrophysicists to solve the mathematical equations that describe the many interactions which take place in a star and allowed the observational astronomers to do massive data processing. Without such computers, it would be difficult for astrophysicists to check their theories against observations or for astronomers to process the data from their instruments.

According to astrophysicists, stars are spheres of hot gas, mostly hydrogen and helium, held together by gravity. Since gravitational forces increase with mass, the larger the mass of the star, the greater the force tending to collapse the star. We can make an analogy between a star and an Olympic weight lifter who sweats and grunts as he holds the barbell over his head, pushing against the force of gravity. As gravity is tending to collapse the weight on the weight lifter, he is exerting an equal and opposite pressure to prevent the collapse. In the case of the weight lifter, that resisting pressure has its ultimate origin in the chemical energy being released in his muscles. But in the case of a star, with its far greater weight to support, where does that opposing pressure come from?

Pressure, like the pressure of a gas inside a balloon, is the result of rapidly moving gas particles colliding with each other or striking a wall. The more rapidly they strike, the greater the pressure; the greater the pressure, the more the gas expands, preventing the collapse of the balloon. The speed of a particle is related to its energy of motion. So the problem of finding a source for the pressure that opposes gravitational collapse in a star is the same as finding the source of heat energy that causes those rapid collisions.

Under normal conditions in a piece of matter, the nuclei of the atoms—their tiny massive cores—are far apart from each other. But the center of a star hardly qualifies as a "normal condition." Indeed, the enormous weight due to the entire mass of the star upon the core —equivalent to about two million tons resting on an area the size of a dime—squeezes the nuclei of the atoms of hydrogen closely together. Besides being under extreme pressure, the core of a star has a sufficiently high temperature to ignite the thermonuclear burning process of hydrogen fusing into helium, a process that generates heat energy. If the temperature in the core of a star is insufficient to ignite the nuclear burning process, a star will not live long—a mere 20 million years.

As a star contracts, about one-half of the gravitational energy released becomes heat energy, which in turn supports the star. Hence for the "short term," gravitational contraction supplies the heat energy. The nuclear burning in the core is crucial only because the heat it generates can compensate the heat loss from the surface of the star, and this halts the contraction for a much longer time. We see that it is the dynamic balance between gravity, which is attempting to collapse the star, and the heat this collapse generates that is responsible for a star's temporary stability.

Stars are not really stable; they only seem so because they live so long compared with us. From their birth out of cosmic gas to their death, their cores are continuously shrinking. To prevent utter collapse, a star must always find new sources of energy that give it an extended lease on life. Chemical sources of energy can keep a star going for only about 20 million years—long compared with a human lifetime but short in cosmological time. The nuclear burning of hydrogen can keep a solar-mass star going for billions of years, and the burning of other elements like helium can extend this period. During the period of nuclear burning stars seem stable, but in fact they are still contracting, albeit very slowly. Ultimately, stars must die because of the relentless crush of gravity and the finiteness of any source of energy.

How can we visualize stars? We can imagine them to consist of a series of layers. Deep in the interior, at the center of a star, is a tiny core, only one-hundredth the size of the full star. The core, consisting of convective currents of hot matter, holds the key to a star's life. Not only do the nuclear reactions in the core provide the heat that prolongs the life of the star, but in the later stages of the life of a star, they also cook up new heavy elements essential to the building of planets and life. Even our bodies are made of star stuff.

Outside the core of some stars is a layer of gas which transfers the radiant energy from the core to the star's outer convective layers. This process resembles the heating of water on a stove, except that it goes on for billions of years. The nuclear burning in the core is like the flame below the pot. The water in the pot transfers this heat to its surface, where the water vaporizes into steam. On the surface of a star like our sun, solar astronomers find a variety of complex physical

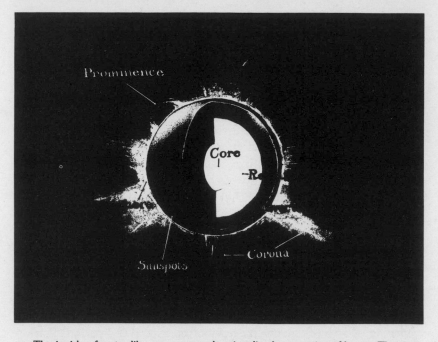

The inside of a star like our sun can be visualized as a series of layers. The tiny core at the center is the location of the nuclear burning process. The layer surrounding the core transfers radiation to the outer convective layers of the star, which then take heat energy to the surface. Sun spots on the surface are manifestations of large magnetic fields. A solar corona of hot bright gas, visible during eclipses, surrounds the sun.

processes not all of which have been understood. But we know the sun has an "atmosphere" extending far out beyond its surface and spilling solar material out into deep space by means of the solar wind. In a sense, the nuclear processes deep within the core of a star are intimately connected with regions of deep space.

The same nuclear processes in the core which produce the energy and pressure to counteract gravity also produce photons—particles of light. But a star is opaque and photons cannot shine directly out of its core. Instead, they randomly bounce around in the interior of the star in a drunkard's walk, colliding off atoms of gas. In about 10 million years, photons produced in the core of a star finally diffuse to the surface—free at last—then travel across interstellar space and eventually reach our eyes.

Today, we understand the complex processes in the interior of stars, but in the nineteenth century, before physicists knew of the existence of nuclear forces, they had difficulty understanding how stars like the sun could release so much energy for such long periods of time. Hermann von Helmholtz, the nineteenth-century German physicist, thought the sun got its energy from gravitational contraction. Lord Kelvin, the English physicist after whom the absolute temperature scale is named, took up Helmholtz' suggestion and calculated the sun's age to be a mere 20 million years. This short life for the sun as calculated by physicists created a terrible puzzle for nineteenth-century geologists and biologists. From the evidence of fossils buried in sedimentary rock layers, they concluded the earth was much, much older than 20 million years. How could the earth and life on earth be older than the calculated age of the sun? That is impossible.

This conflict over the ages of the sun and the earth created a division among nineteenth-century scientists into "catastrophists," who believed that God periodically intervened in nature (by, for example, flooding the earth), and "uniformitarians," who believed the world evolved slowly over long periods of time, guided by natural laws. Lord Kelvin, who vigorously defended his calculation of the age of the sun, led the attack on the uniformitarian view. Today, of course, assuming that nuclear energy is the source of the sun's energy, twentieth-century astrophysicists have shown that the sun's age is comparable to the 4.5 billion years now estimated for the age of the earth. It is a gratifying sequel to the nineteenth-century dispute among physicists, geologists and biologists that today they all agree on the chronology for the earth's history—a chronology measured in billions, not millions, of years.

Stars live a long time. According to astrophysicists, the lifetime of a star is roughly proportional to the inverse of its mass squared (more generally, an inverse-power law). Hence a star ten times as massive as the sun lives only one-hundredth as long as the estimated 10 billion years of our sun—a mere 100 million years. A star ninety times more massive than the sun can live for only a million years—nothing on the cosmic time scale. This may explain why we do not see many such supermassive stars—they disappear very quickly. If stars get more massive than about ninety times the sun's mass, then the crushing weight of the star heats up the core to very high temperatures and blows away the outer layers of the star, reducing its mass. Such supermassive stars are thus not stable, so that ninety solar masses seems to be a maximum mass for a star.

What about the minimum mass of stars—stars much less massive than the sun? Such stars are hard to see because low-mass stars are not very hot or very bright. Observational astronomers have a kind of informal contest to find the least luminous star—a star that is intrinsically dim, not just dim because it is very far away. For a while, the record was held by the star VB10 in the constellation Aquila, but this star was recently overtaken by star RG0050-2722 in the constellation Sculptor.

Looking for the least luminous star is not an idle pastime, because knowing the lowest stellar luminosity has considerable importance for the theory of stars. Since the luminosity of a star is related to its mass, the least luminous star also has the lowest mass—about 2.3 percent of the sun's mass for the current record holder. Below a certain mass, stars will not ignite their nuclear furnaces and cannot burn. Thus, knowing the lowest mass for a star would provide an important constraint on theoretical models of the star-forming process. Without sufficient mass, the gas out of which stars are formed cannot concentrate sufficiently to make a star, and it is important to know what the minimum mass is.

One question that arises if we contemplate low-mass astronomical objects is Where do stars end and planets begin? Jupiter, the giant planet of our solar system, has a mass of only 0.1 percent of the sun's mass—about one-twentieth the mass of the lowest-mass star. Astrophysicists believe that the process of making planets, even big planets like Jupiter, is different from the star-making process. Planets, according to the most widely held theory, are supposed to form out of a flattened disk of gas and cosmic debris surrounding a newly born star. Astrophysicists suspect that there may not exist objects with masses between that of the least luminous star and that of large planets like Jupiter; or perhaps they just haven't found them yet.

As stars burn away their nuclear fuel, they continually make adjustments, like the adjustments parents make in clothes for a growing child. Because of these adjustments, shifts and movements, stars such as our sun are making noise—they have a whole symphony of sounds. What is the origin of the solar song?

Recall that the light created by nuclear burning in the core of a star cannot get out right away because the stars are opaque. Consequently, the light radiation heats up the gas in the star's outer layers, stirring them up the way the sun heats the air on a hot day. The hot gases carry their heat to the surface of the star. All this interior movement of gases inside a star creates sound waves. These sound waves bounce around inside the star (it takes one hour for sound to cross the interior of our sun), which acts like a giant loudspeaker.

According to some astrophysicists, in our sun this acoustic energy is dumped into the solar corona—the upper, very hot atmosphere of the sun. Others, contesting this view, think that the acoustic energy is dumped into the sun's chromosphere, its upper layer. The corona is instead heated by electrical currents generated by the solar magnetic field. Because the solar corona is not very dense, it cannot radiate away any extra energy and instead expands, carrying away the energy like a powerful jet engine blowing away hot gases. This expanding solar corona is called the "solar wind," and it stretches out far beyond the earth to the outer planets. As part of the cosmic recycling system, the solar wind dumps hundreds of millions of tons of solar material into outer space each second. By using artificial satellites that can move through the solar wind and transmit back data about its activity, scientists hear the creaks, groans, screams, thunderclaps and drumrolls of our sun's song.

The sun not only sings, it vibrates. By carefully observing the shape of the sun, scientists observe that it is vibrating in various frequency modes like a shaking bowl of gelatin. Some of these vibrating modes are as short as minutes; others take hours. Such observations of the complicated external movement of the sun give scientists clues about the internal motion of the core, which they cannot directly see. For example, by analyzing the external modes of vibration of the sun, scientists can determine how fast the core is rotating relative to the outer layer.

As these discoveries indicate, scientists have learned a great deal about stars by a close study of our local star, the sun. Today we are living in the golden age of solar astronomy. Almost yearly new and interesting features of the sun are uncovered. More will be learned in the coming decade as artificial satellites designed to examine the sun close up are sent on their maiden journeys. Yet the sun is only one star out of billions. If we are to grasp the variety of stars, their types and evolution, the study of just the sun is insufficient. How do astronomers study the stars that are far away?

It is sobering to realize that almost all the information astronomers acquire about the faraway parts of the universe comes to them via

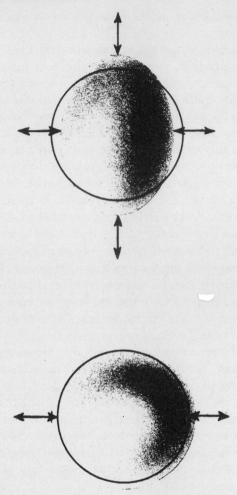

Simple vibrational modes of the sun. Solar astronomers observe such vibrations, as well as more complex ones, for clues to processes occurring deep in the interior.

electromagnetic radiation—light in various wavelengths, corresponding to visible light, radio waves, infrared and microwave radiation, X rays and gamma rays. By studying the message of light from distant sources, astronomers put together a picture of the universe. Light carries an enormous amount of information, as is apparent if we reflect on how much of our knowledge is based upon what we see.

Two important pieces of information that starlight carries are the color of the star, ranging from reddish to bluish, and the luminosity of the star. The color is related to the surface temperature of the star, with bluish color implying a very hot surface, while the red-tinted stars are relatively cool. The luminosity of the star corresponds to the total energy output of a star. If one plots the luminosity versus the temperature of lots of different stars on a two-dimensional graph, one finds not a random scatter of points but a narrow region, a line, along which most of the stars lie rather nicely. This graph is called a Hertzsprung-Russell diagram (after its discoverers), and the line running from the hot, bright stars to the cool, dim ones is called the "main sequence."

Knowing the location of a particular star on the Hertzsprung-Russell diagram is perhaps the single most important piece of information astronomers have about a star. It tells them how massive the star is, a fact that determines its fate as well as many other properties. As stars evolve by burning hydrogen, they move slowly along the main-sequence line, eventually coming to that point in their lives when dramatic events occur and they move off the main sequence. Those stars which do not lie on the main sequence at all are known to be special in some way. Examples are the red giants and the white dwarfs.

Although the luminosity and temperature are very important characteristics of stars, perhaps the most detailed knowledge of a star comes from observation of the spectrum of a star's light. Starlight appears white, but if this light is refracted through a prism it is separated into a spectrum of colors. Besides a possible continuum of colors in the spectrum ranging from red to blue, there appear distinct spectral lines of intense, narrow bands of color. Other lines of color may be missing. These spectral lines of specific colors, or wavelengths, of light are due to specific energy changes in the atoms in the star which emit the light. From the pattern of colors in a star's spectrum, we can deduce the kinds of atoms—the chemical elements—in a distant star. Stellar spectra are the fingerprints of stars.

In 1825, Auguste Comte, the French social theorist and founder of positivism, remarked in his *Cours de Philosophie Positive* that one thing we would never know is the chemical composition of stars. At the very time he made this remark the Germans Joseph von Fraunhofer

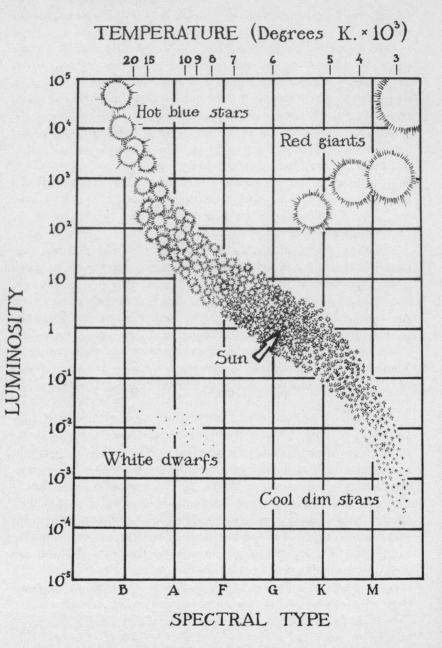

TEMPERATURE (Degrees K. × 10³)

Hot blue stars

Red giants

LUMINOSITY

Sun

White dwarfs

Cool dim stars

SPECTRAL TYPE

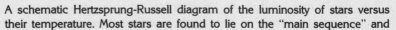

A schematic Hertzsprung-Russell diagram of the luminosity of stars versus their temperature. Most stars are found to lie on the "main sequence" and

and later Gustav Kirchhoff and Robert Bunsen were laying the foundations of modern spectroscopy. A few decades later, the study of the chemical composition of stars was an exciting scientific field. Comte's remark reminds me of those people, scientists included, who today say that one thing we will never know about is the very origin of the universe. Just as Comte thought we would never know about the composition of stars because they are so far away in space, people today argue that we will never know about the origin of the universe because it is so distant in time. But they are mistaken.

By examining their spectra, scientists found that stars differ in their chemical composition. All stars contain mostly hydrogen and helium. But some have a relatively large abundance—1 to 2 percent—of chemical elements heavier than hydrogen or helium, while others have only a trace—a tiny fraction of 1 percent—of these heavy elements. This distinction leads to a classification of stars into population I—those with heavy elements—and population II—those relatively lacking in heavy elements.

Walter Baade, the German-American astronomer, discovered these two distinct populations of stars in 1942, during World War II, using the 100-inch telescope at Mount Wilson, near Los Angeles. Interestingly, the war played an indirect role in this discovery. Because Baade was a German national and not able to join the American war effort, he was left as the sole user of the telescope. Furthermore, because of the possibility of enemy attack Los Angeles was blacked out at night, and the absence of city lights created excellent observing conditions at the Mount Wilson observatory. Baade was able to resolve many individual stars in the central bulge of the Andromeda galaxy, and there he found the two different populations of stars. Population I stars were the young bluish ones, while population II were older red stars. Outside the central bulge, in the spiral arms of the galaxy, he found mostly young population I stars like our sun. The globular clusters of stars that inhabit the halo surrounding a galaxy date back to the time of its formation, and these consist of the older population II stars.

What is the origin of the two populations of stars? Most astrophys-

move along it in the course of their lives. The location of our sun is indicated by the arrow. In the upper right are the red giant stars, while in the lower left are the white dwarfs—both special stars because they do not lie on the main sequence.

icists believe that only light elements like hydrogen and helium, with a bit of deuterium and lithium but no heavy chemical elements, got cooked up in the big bang that was the origin of the universe. Astrophysicists call these first elements "primordial elements" to distinguish them from elements that later got synthesized in the nuclear furnaces within stars or in supernova explosions. In the beginning, the universe was just a gas of primordial hydrogen and helium. Stars came later when the gas condensed to form them. The first stars cannot have had any heavy elements at all because there simply were not any around. These "pure" stars—hypothetical stars because no one has ever seen one—are called population III stars. But these primordial stars did important work cooking up the first heavier elements like carbon and nitrogen from the light ones like hydrogen and helium by nuclear burning and transmutation. When the population III stars died in spectacular explosions, they spilled out their production of heavy elements into interstellar space. These rare heavy elements eventually ended up in the oldest observed stars—the population II stars. These, in turn, cooked up more heavy elements, which were spilled out into space and ended up in the younger population I stars. Such heavy elements in the interstellar medium of gas and dust which pervades the space between stars, tend to stick together when the atoms collide, making up little dust grains one-hundred-thousandth of a centimeter in size. These tiny dust grains congregate in the dark, nebulous clouds in the spiral arms of the galaxy which are the birthplaces of new stars. By a grand recycling process new stars are made from the material of dead ones. But how exactly are stars born?

It is unlikely that we will ever see a star being born. Stars are like animals in the wild. We may see the very young but never their actual birth, which is a veiled and secret event. Stars are born inside thick clouds of dust and gas in the spiral arms of the galaxy, so thick that visible light cannot penetrate them. Yet in spite of the impossibility of directly observing the birth of stars, astronomers have made great strides in understanding this complex process. Many puzzles remain because the birth of stars involves so many kinds of physical interactions, but most astrophysicists are confident that a complete theory is on the way.

The galaxy is a dirty place. In the space between the stars, there is an interstellar medium consisting of gas (single molecules and atoms) and dust (tiny specks of matter) flying about. The interstellar medium is very thin—the whole medium is no more than 5 percent of the mass of the galaxy—but it gets concentrated in particular places. Our understanding of the interstellar medium was revolutionized by the *Copernicus* satellite launched in 1972 and operated until 1980. This automatic satellite was designed to measure the spectrum of light in the blue, far ultraviolet region, light that cannot penetrate the earth's atmosphere and can only be detected above it. From data collected by this satellite, astronomers learned about the abundance of various atoms and molecules in the interstellar medium, their temperature and many other features.

The interstellar medium pervades the disk of the galaxy. It is especially dense in the spiral arms delineated by bright young stars. The relation between the spiral arms and the bright young stars found in them is like the relation between the chicken and the egg: Which came first? Do huge waves in the interstellar medium randomly sweep material into the spiral arms, thereby creating the perfect conditions for the formation of young stars? Or is the existence of stellar "nurseries" a precondition for the formation of the arms? Theoretical astronomers struggle with such difficult questions and devise computer models of the star-formation process in order to answer them. But independently of any such computer models, it is clear from observations alone that gas and dust concentrate in the form of thousands of "giant molecular-cloud complexes"—huge gas clouds—and that is where stars are born.

As recently as fifteen years ago, astronomers thought that most of the gas in our galaxy consisted of single atoms. But using radio telescopes and satellite detectors that were sensitive to radiation emitted by molecules, they found that 10 to 50 percent of the gas in our galaxy is molecular—atoms stuck together—and that it tends to cluster in giant clouds. About 99 percent of this gas is molecular hydrogen—two hydrogen atoms stuck together—but at least 53 other molecules have been detected, including ethyl alcohol, or vodka. The clouds of gas near the center of our galaxy contain enough vodka to fill more than 10,000 goblets the size of earth.

One property of molecular (in contrast to atomic) clouds of gas is that they are much colder and denser. These giant molecular-cloud complexes are the most massive objects in the galaxy. Astronomers have detected more than 4,000 of them, mostly populating the spiral arms. Continually changing in complicated ways, the clouds seem to live only a short time, a hundredth of the age of the sun or earth, and astronomers speculate about their birth, middle age and death. But it seems clear that the clouds are the birthplaces of very massive, short-lived stars and possibly long-lived stars as well.

Looking at the constellation of Orion on a clear night, you just might see the great nebula in the Hunter's sword not far from the three bright stars in the Belt. The Orion nebula is a stellar nursery rich in complex physical phenomena, a "laboratory" in the sky and the site of giant molecular-cloud complexes. At the center of the huge clouds of churning dust and gas is the Trapezium, a set of four bright stars that act upon the nebular material, illuminating it while exciting the gas to glow on its own. Huge shock waves, which perhaps trigger star formation, can be seen propagating in the clouds. New stars have been made deep inside the Orion nebula where we cannot see them. How do astronomers know that?

More than 180 years ago, William Herschel noticed the existence of "radiant energy" beyond the red end of the spectrum of visible light. In modern terms he found that the sun emits not only visible light but also light in the long wavelength, infrared region. Today, infrared astronomy is a major new branch of astronomical science, with a few observatories devoted to it already built. These include the NASA Infrared Telescope Facility (IRTF) and the United Kingdom Infrared Telescope (UKIRT), both on top of Hawaii's Mauna Kea, an extinct volcano; the Wyoming Infrared Telescope; the Multiple Mirror Telescope (MMT) in Arizona; Mexico's new 2.12-meter reflector and most recently the one-ton *Infra-Red Astronomy Satellite (IRAS)* placed into earth orbit but now no longer operating. With such instruments sensitive to infrared light, a new and exciting understanding of previously invisible components of the cosmos is now revealed.

Unlike visible light, infrared radiation can penetrate the thick clouds of the stellar nurseries like the Orion nebula. By detecting microwave and infrared radiation, two astronomers, Eric E. Becklin and Gerry Neugebauer of the California Institute of Technology, in 1965 discovered a mysterious object—called the B-N Object—deep inside one of Orion's clouds. In the following decade about six other diffuse centers of infrared-radiation emission were detected in the giant cloud. At one time astronomers speculated that these were protostars—objects that were on the verge of becoming true stars which burn nuclear fuel. But now it appears that these infrared sources are due to stars further along in their lives than protostars—

young massive stars. These young stars have extremely energetic outflows of mass, a gigantic stellar wind of particles whose origin is unknown. At about a light-month's distance from the star, the stream hits the surrounding gas cloud, heating it and producing microwave and infrared radiation. The gas cloud actually amplifies the microwave radio signals, and these are detected here on earth—a beautiful confirmation of the physical processes involved. Infrared and radio astronomers discovered similar objects—young stars still encased in cocoons of dust and gas—not only in the Orion nebula but also in the Omega and Trifid nebulae—perhaps the most convincing evidence that these nebulae are star-forming regions.

A major boost for infrared astronomy was the January 25, 1983, launching of *IRAS*, which made a complete infrared map of the sky. Every seven months its telescope swept the entire sky twice. Its principal mission was to examine the hot spots where astronomers suspect stars are made. The satellite spotted dozens of infrared sources in the Tarantula nebula located in the Large Magellanic Cloud, a neighboring galaxy. Some of these sources are probably due to young stars' interacting with the thick dust surrounding them; others may be true protostars—a possibility that has astronomers excited.

How is a star made? To answer this, we turn to astrophysical theorists and their computer models. They construct a picture of the formation of stars in giant molecular-cloud complexes—clouds consisting mostly of molecular hydrogen and whose mass is far greater than that of an individual star. At first, the large cloud begins to fragment into smaller clumps, a result of the mutual gravitational attraction of all the gas and dust particles, and the mass of these individual clumps is about the mass of a star. So we learn that some stars are born in groups. Astronomers have seen many "open clusters," groups of young stars all moving together in the same direction and which were probably born together out of the same concentration of dust and gas.

Massive, short-lived stars get made in gigantic gas clouds; others, shunning such mass-production systems, are born individually out of a different class of much smaller dark clouds known as "globules." About two hundred of these nearly spherical globules, rich in large molecules, have been found within 15,000 light-years of the sun, and some of them are collapsing with just the right amount of mass to form a single star. Conceivably our sun was created from such a globule. The existence of the dark globules indicates that there may be many other ways that stars are made besides in giant gas clouds. And even within a giant cloud, there may be different mechanisms at work that produce stars. But let us examine what happens to a clump of gas in a star-forming region.

Following the progress of an individual clump, we see complex physical processes at play. Grains of dust, although they make up only about 1 percent of the mass of the clump, are important sites on which the surrounding gas molecules can stick and form. The presence of dust may also protect the molecules from intense radiation that might dissociate them. The ubiquitous force of gravity is trying to pull the dust-and-gas clump together while the action of magnetic fields, heat, turbulence and spin in the clump tends to disperse it, competing with gravity. In the end, gravity wins all such competitions. In the case of the formation of a star, this competition is a long process—about 10 million years. Gravitational contraction is accompanied by increasing density of the clump, and when it becomes opaque its temperature rises. By the time the clump is some hundred times the size of our solar system, it has reached a temperature of zero degrees Celsius—the freezing point of water. When the clump has contracted to the size of the solar system, its temperature is thousands of degrees Celsius—hotter than the melting point of metals. About 100,000 years after it began contracting, the clump could fit into the earth's orbit and its temperature is hundreds of thousands of degrees. Such an object, which is not yet a star, is called a "protostar."

If we follow the progress of a single protostar, which lasts about 10 million years, it continues to contract and get hotter. Finally, the temperature in the center reaches the 10 million degrees Celsius required to ignite the hydrogen-fusion nuclear reaction, and a true star is born.

OPPOSITE:
A schematic representation of the star-formation process out of a gravitationally contracting cloud of gas and dust. A giant cloud located in a spiral arm of our galaxy fragments into lumps, each of which then fragments further. Such a spinning, individual lump continues contracting over a period of millions of years. Here it is shown forming a binary star system—one possible outcome. Eventually, the stars ignite and blow away the remaining clouds. Such star births are accompanied by jets of matter ejected from the new stars, whose origin is not yet understood.

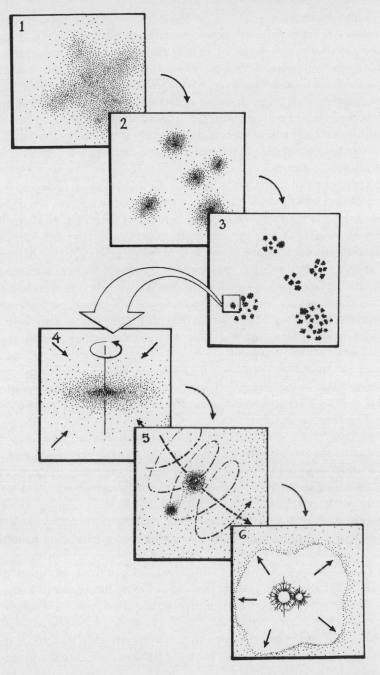

This simple theoretical picture of a collapsing cloud presents problems if compared with observations. Astronomers indirectly observe two fast-moving streams of matter flowing in opposite directions from newborn stars; these streams are millions of times more intense than the solar wind blowing away from the sun. The picture of continuous contraction does not account for the origin of the mysterious twin jets of matter emitted by young stars. Astrophysicists now realize that the birth of stars is dynamically more complex than they previously thought and are hard at work devising new models and modifying old ones.

Some computer models of the star-formation process currently in vogue find two possible outcomes for a contracting cloud, depending on, among other things, the initial amount of rotational momentum. One possibility is that the contracting dust and gas forms into two orbiting stars—a binary system. The other possibility is that only one star forms, with the rotational momentum distributed into an auxiliary planetary system. If these models are correct, then half the stars in our galaxy, since they are observed to be binaries, would not have accompanying planets while the other half—single stars—probably have planetary systems. This strong conclusion—that all single stars have planets—depends on complicated details of the rotational-momentum-transfer process within the gas cloud, which is not well understood. Some astrophysicists think that the rotational momentum is carried off by a stellar wind and that the rotational momentum of the planets is unimportant.

In spite of these complications, many astrophysical theorists believe that the birth of a single star like our sun was accompanied by a disk of leftover matter swirling around it like the white of a fried egg surrounding the yolk. This disk is called the "solar nebula," and from it the solar system of planets subsequently formed. But the mathematical problems in making theoretical models of the solar nebula are especially difficult because there are no observational data to guide the model builder. No one has seen a planetary system in any stage of formation, except the last. But if we believe these models of the solar nebula, then the planets formed as the matter in the disk began to thicken into lumps. Larger lumps accumulated yet more pieces of matter. One theory holds that the asteroid belt between Mars and Jupiter is a collection of lumps that never combined to build a true planet. When the sun ignited for the first time, it emitted a gigantic wind which blew away all the debris that did not form into massive bodies like planets and moons.

If this scenario for the origin of the solar system is correct, then planetary systems ought to be at least as common as stars like our sun, a typical single star. From data already obtained from probes sent to the planets, we will learn much more about the origin of our solar system in the coming years. But no quick answers should be expected. In spite of manned landings on the moon, moon rocks taken to earth and selenological data, we still are uncertain about the origin of the moon—let alone the origin of the solar system.

The theoretical study of the solar nebula, the flat disk of matter surrounding the nascent sun, is but a specific example of the general astrophysical phenomenon of "accretion disks." The rings around the planet Saturn are another example of such an accretion disk, and so, perhaps, are the stars in a disk galaxy. Black holes and neutron stars also ought to be surrounded by a disk of hot gas, and it seems likely that an energetic signal is emitted when the gas falls into the hole or star. In special circumstances, matter such as gas or dust evidently tends to form a semistable disk surrounding a massive object. The mathematical study of accretion disks will eventually lead to a deeper understanding not only of the origin of the solar system but also of the puzzling signals that emanate from deep space.

About half of all stars near us go through life with a companion; they are paired together. A few, disdaining such stellar social conventions, are members of triplets or quadruplets. The binary pairs have been extensively studied. Some binary pairs orbit each other so closely that they are classified as "contact binaries"—they effectively touch each other, exchanging large quantities of mass. For many years astrophysicists were puzzled because members of binary pairs appeared to have very different ages. They reasoned that if they were born at the same time out of the same gas cloud then they should have the same age. But if the two stars exchange mass during their evolution, then their apparent age can be quite different, since adding mass to or subtracting it from a star can change its apparent age. The observation of contact binaries along with computer modeling confirmed these exchange mechanisms.

Most stars, once formed, lead uneventful, middle-class lives. Burning away hydrogen in their cores, cooking up helium, making adjustments, they sing and vibrate for billions of years. A very small fraction of stars exhibit quite unconventional behavior. Among these are the T-Tauri stars, wild baby stars, only a hundred thousand to a million years old—very young for a star. The T-Tauri stars exhibit rich, complex, often anomalous emission spectra. They are usually surrounded by hot gas and luminous jets of matter, all of which adds to the difficulty of interpreting the physical processes. T-Tauri stars are puzzling because of the diversity of peculiar features they exhibit, a diversity which, once understood, will teach astrophysicists a lot about these strange, youthful stars and the gases that accompany them.

Another class of interesting stars are the Cepheid variables—variable because their brightness oscillates with a period ranging from three days to weeks. They are blinking beacons in the sky. Polaris—the North Star—is a Cepheid variable and alters its brightness by 10 percent every four days. Cepheids are old stars which have discovered that they can release their energy more efficiently by pulsating their intrinsic brightness. But the most remarkable property of a Cepheid is that its rate of blinking is precisely related to its brightness.

This important relation—the period–luminosity relation—was discovered in 1912 by Henrietta Leavitt of the Harvard College Observatory while she was examining photographic plates of the Small Magellanic Cloud sent to Harvard from an observatory in Arequipa, Peru. Since the Magellanic Clouds, irregular galaxies on the border of our Milky Way galaxy, are very far away, their stars are all approximately the same distance (just as Paris and Lyon are approximately the same distance from San Francisco). Because we know that all these stars are at the same distance from us, we may conclude that the *apparent* luminosity (the brightness that we see in individual stars through a telescope) is proportional to their *intrinsic* luminosity (the total amount of energy in the form of light leaving the star). Leavitt noticed that the periods of the Cepheid variables in the Small Magellanic Cloud were related to their apparent luminosity and hence similarly related to their intrinsic luminosity as well. What she found was that the brighter the star, the longer is its pulsation period, and this became known as the period–luminosity relation. By calibrating this relation, using nearby Cepheids (to which we know the absolute distance), we can obtain a relation between the observable period and the apparent luminosity on one hand, and the distance to the star on the other hand. The importance of the Cepheid variables is that they provide astronomers with a method for obtaining the distance to faraway stars and even the distances to other galaxies in which we can see individual Cepheids. The blinking Cepheids are the rulers with which the size of the universe is measured.

Who could have imagined what we now know of the birth and life of stars even a century ago? We have lived in a golden age of astrophysics in which the basic processes for the life of stars were first understood. There remain problems. But these are not problems of principle but rather problems of complexity, a complexity that will continue to engage astrophysicists in the decades to come. Aided by new instruments and computers, astrophysicists will continue to build models of stars, testing their theories against increasingly refined observation.

Ancient people worshiped the sun as the source of life. In the future, as our knowledge of stellar systems grows, we may learn that planetary life is part of the evolution of a star system. Stars create the conditions for life and so we are bonded to them. But the stars themselves are not eternal, independent beings; they are the progeny of the galaxy. Their life, and hence ours, is intertwined with galactic processes occurring on time scales that are incomprehensible from a human perspective. Yet, viewed from a great temporal perspective, no part of the universe is truly independent of the whole. Is it not possible that just as the environment on earth has shaped life here on time scales of millions of years, the environment of the universe will shape the future of life on time scales of billions of years? Life may find that the whole universe becomes the stage of its existence. Is our destiny among the stars? Or is such a starry vision of our future but an illusion reflected endlessly in the mirrors of our mind?

3

The Death of Stars: Astronecroscopy

This "shuddering before the beautiful," this incredible fact that a discovery motivated by a search after the beautiful in mathematics should find its exact replica in Nature, persuades me to say that beauty is that to which the human mind responds at its deepest and most profound.
—Subrahmanyan Chandrasekhar

Stars are an image of eternity. They appear fixed, immutable and indestructible, and astronomers thought of stars this way for centuries. But within the lifetime of people living today this view of the eternity of stars has been drastically changed. Like living things, stars die. Their substance is transformed, their elements are scattered about the galaxy and the remains of some of them become sealed in celestial tombs so secure that they are beyond the reach of time and space.

Understanding the fate of stars is part of a more general puzzle in physics called the "final-state problem"—the problem of determining what ultimately happens to anything whatsoever if we wait long enough. Many of the material things that surround us seem to have a quality of permanence—the seas, mountains and atmosphere of the earth. Yet if we wait long enough all material things are transformed. Even the atoms out of which they are made are fated for extinction and annihilation. Where do things eventually end up? What is the fate of our galaxy and the universe? Physicists speculate on the answer to this question and come to a variety of conclusions—the issue is not settled. But some definite conclusions can be reached if we turn to examining the fate of stars. There we discover new physical processes which so surprised even the scientists who first understood them that they only reluctantly accepted the conclusions of their reasoning. For the crushing force of gravity that accompanies the violent death of stars creates physical conditions which challenge our very understanding of the laws of nature. Let us follow a star in its death agony.

After billions of years of burning hydrogen and converting it to helium ash, a star runs out of hydrogen fuel in its core—an energy crisis that seals its fate. Recall that nuclear burning provides the sustained high temperature for resisting gravitational collapse. When that burning stops, the star resumes collapsing. Astrophysicists envision three possible fates for collapsing stars: they become either white dwarfs, neutron stars or black holes. Which of these three fates lies in store for a particular star depends primarily on its total mass. Stars less massive than 1.4 times the sun's mass become white dwarfs —tiny stars made of matter thousands of times denser than ordinary matter. More massive stars undergo a supernova explosion with a neutron-star remnant—essentially a gigantic atomic nucleus the size of a city. Stars with a core mass larger than about 2 solar masses are expected to collapse into a black hole—an object in which space itself gets turned "inside out." How did astronomers come to these bizarre conclusions about the death of stars? What evidence is there for such strange objects? Let us examine each in turn.

WHITE DWARFS

Shut up. Don't talk nonsense.

—Arthur S. Eddington

No one in modern times has ever seen a star in our galaxy collapse; but astrophysicists can construct a picture of the final years of a star through the use of computer models that implement the laws of nuclear physics and thermodynamics. Once the hydrogen in the core of a star has turned into helium, the star's precious equilibrium is lost, and the tiny core (only about one-hundredth the size of the whole star) begins to compress under the immense pressure of the outer gas layers of the star. This gravitational compression heats the core, which in turn heats the outer layers, causing the hydrogen in these layers (which is not yet depleted) to burn ferociously.

Through a complex interplay of energy-transfer processes, the outer layers expand, ballooning the surface of the star way out. The star is now swollen to thousands of times its former volume and turns red —a reflection of the fact that the outer layers, by expanding, have cooled, and the gas at the lower temperature is red instead of white. Such a star—a hot tiny core surrounded by a huge envelope of hot gas—is called a red giant. Examples are the red stars Pollux and Arcturus.

The core continues to compress to a density a thousand times the density of the core in a normal star until the rising temperature at the center of the core reaches 100 million Kelvin. At that high temperature a new burning process is initiated. The helium nuclei in the very center of the core fuse together to form the heavier element carbon, and in this process the whole star undergoes a reduction of its huge volume to achieve a new equilibrium. The core, with its newfound energy derived from the burning of helium into carbon, has given the star a new lease on life. But not for long.

What happens next depends primarily on the total mass of the star. Either the outer layers of the star are sufficiently massive to continue compressing the core and heating it still further or they are not. High-mass stars turn into neutron stars or black holes. In relatively low-mass stars like our sun, the outer layers just do not have the weight to keep compressing the core. Instead, the intense heat generated by the helium burning in the core literally blows away the outer layers of the star into interstellar space. The resulting filaments of hydrogen gas are called "planetary nebulae," and about a thousand have been spotted in our galaxy. The term "planetary nebula" is a misnomer committed by the first astronomers who observed the nebulae and thought they looked like planets. These nebulae have nothing to do with planets; they are the remains of the outer envelope of dying stars.

Eventually the filaments of gas disperse into interstellar space and all that remains of the star is the naked core, about the size of the earth—a white dwarf star. For eons, the white dwarf loses energy, turning from white to yellow to brown and finally to black. The transition from white dwarf to black dwarf takes so long that no black dwarfs may yet exist in our galaxy. But what about white dwarfs? Do they exist?

The story of white dwarfs begins in 1844 at the observatory in Königsberg (then part of Prussia, now part of the Soviet Union), when Friedrich W. Bessel saw that the image of Sirius, the brightest star in our heavens, wobbled. What causes the image of a star to wobble? Bessel concluded that Sirius was accompanied by a massive dark companion star which, as it orbited, pulled on Sirius, causing a wavy motion in Sirius' position in the sky. Bessel did not see the dark star, but his guess proved correct when nineteen years later Alvan Clark, an American telescope builder, spotted the dim companion of Sirius while testing a new 18-inch lens. Alvan Clark was a member of a distinguished family of American telescope builders who later made the large refracting lens for the first telescopes at the Lick and Yerkes Observatories.

But there was something odd about the companion of Sirius. In 1910, Henry Norris Russell, the codiscoverer of the Hertzsprung-Russell diagram, noticed that this star did not fit on the main sequence, and that exception caused him a lot of worry. Maybe the correlation he had found between the surface brightness and density of stars was all wrong. He asked the astronomer Edward Pickering to get the spectrum of Sirius' companion for him. Russell reported:

> Characteristically, he sent a note to his observatory office and before long the answer came . . . that the spectrum of this star was A. I knew enough about it, even in those paleozoic days, to realize at once that there was an extreme inconsistency between what we then would have called "possible" values of the surface brightness and density. I must have shown that I was not only puzzled but crestfallen at this exception to what looked like a very pretty rule of stellar characteristics; but Pickering smiled upon me and said, "It is just these exceptions that lead to an advance in our knowledge."

In the next seven years, two more such exceptional stars were discovered.

Normally, dim stars (this one was truly dim—only one four-hundredth the intensity of the sun) should have a red color, while Sirius' companion was instead burning white hot. The only explanation for its dimness was that it was extremely small. But if it was so small, then it would not be sufficiently massive to influence the observed movement of a heavy star like Sirius. A way out of this puzzle was to assume that the companion of Sirius was indeed very small but made of matter three thousand times as dense as the matter in ordinary stars. But that solution to the puzzle seemed like nonsense. No such dense form of matter was known to exist in the beginning decades of this century. Reflecting on this puzzling message from the dim companion of Sirius, the British Astronomer Royal, Sir Arthur Eddington, said in 1927, "What reply can one make to such a message? The reply which most of us made in 1914 was—Shut up. Don't talk nonsense."

The message was indeed nonsense if interpreted in terms of Newtonian physics. The resolution to the puzzle of Sirius' companion had to await the invention of the quantum theory of atoms in 1927 and the work of a nineteen-year-old Indian, Subrahmanyan Chandrasekhar, done in 1930. Building on the earlier work in England of Ralph H. Fowler, who showed that when a star exhausted its nuclear fuel it had to collapse, Chandrasekhar saw what it had to collapse into: a new superdense form of matter, so dense a cubic inch would weigh ten tons. How can we think of such matter?

Fowler had made use of the quantum physicist Wolfgang Pauli's 1925 discovery of the "exclusion principle." According to Pauli's exclusion principle, electrons (which are small electrically charged particles swarming about the atomic nucleus) cannot sit one on top of another—they exclude each other, and if you try to push two electrons in the same state together they will repel each other. This repulsive force is not due to the fact that the like electric charges on the electrons repel each other, but is an entirely new kind of repulsive force far stronger than the electric force. This new force, called an "exchange force," is understood only on the basis of quantum theory and has no analogue in classical physics. Its existence at the atomic level is what keeps the electronic clouds surrounding atoms from collapsing.

If we imagine a gas of electrons and then imagine applying pressure to it, the repulsive exchange force between individual electrons will set up an opposing "Fermi pressure" to resist this squeezing. But you have to press hard on the gas before you feel this resisting Fermi pressure. It comes into play only when the electrons are pushed together so closely that their associated waves begin to overlap. Such conditions exist inside of stars. What Chandrasekhar realized was that the special relativity theory implied that the Fermi electron pressure, born of the weird world of quantum theory, would resist gravitational collapse and stabilize the star, provided its total mass was not too large. He calculated that this would be the case in stars with a mass less than 1.4 times the mass of the sun—a critical mass that is called the "Chandrasekhar limit." In some such stars the density of matter for which the equilibrium between gravity and Fermi pressure is reached is 10 tons per cubic inch—just right to explain the behavior of the companion of Sirius. This star, a white dwarf, was once a normal star, but then it ran out of hydrogen fuel in its core and stabilized again through the occurrence of the Fermi pressure. Today astronomers have detected more than three hundred white dwarfs.

Some white dwarfs are, like Sirius' companion, members of a binary star system in which the other member is a normal star. The dwarf can orbit quite close to the normal star and draw gas from it. The gas, mostly hydrogen, falls onto the dwarf and begins to accumulate, and after a sufficient lapse of time a critical amount is reached. Then, as the hydrogen fuses into helium, it explodes all at once on the surface of the dwarf like thousands of hydrogen bombs. Hundreds of such "nova" explosions have been observed—additional confirmation of the bizarre properties of white dwarfs.

NEUTRON STARS

On a chi-chhou day in the fifth month of the first year of the Chi-Ho reign period [July 4, 1054], a guest star appeared at the south-east of Thien-K'uan, measuring several inches. After more than a year, it faded away.
—Toktagu, *Records of the Sung Dynasty*

A more spectacular fate awaits stars more massive than the Chandrasekhar limit of 1.4 solar masses. In such stars, after the helium begins to burn in the core, the outer layers of the stars have sufficient mass to keep up the pressure on the core so that it continues compressing and, therefore, heating. The temperature rises so high that new nuclear burning processes are started. The carbon core burns furiously and quite rapidly cooks up even heavier elements. The inside of the old star soon exhibits, like an onion, distinct layers. On the outside of the core are the lighter elements hydrogen and helium; in the middle layers one finds carbon and helium; and as one penetrates still deeper, the layers contain successively heavier elements— magnesium, silicon, sulfur, and so on up to iron, the heaviest element that gets made in a star by standard nuclear burning.

The core of the star consists mainly of iron. Iron is not the heaviest element, but it has the special property that it will not undergo nuclear burning. Iron is the final ash of nuclear burning—there is no way to extract energy from iron nuclei by fusing them together. It is not clear what happens next. But it seems that once sufficiently large amounts of iron have been synthesized in the core, the nuclear burning stops, the pressure preventing the gravitational collapse of the star disappears abruptly and the star undergoes catastrophic collapse. The immense mass, prevented from collapsing to the center of the star for billions of years, now does so in a matter of seconds. That energy—equivalent to the energy output of the star over its full previous billion years of life—is released in a few seconds, and the explosion is as brilliant as a billion suns. If one of our neighboring stars underwent such a supernova explosion (none is so fated except possibly Sirius) then a second sun, as bright as our own, would appear in the sky and roast us alive.

What happens during the supernova explosion? Since no one has seen a supernova in our galaxy since 1604, when Kepler's star in the constellation Serpens exploded, we have not had the opportunity to watch one close up with modern instruments. A "supernova watch" has been established among astronomers so that if one does occur in our galaxy (and it is "overdue") many observatories will immediately train their instruments on it. About four hundred supernovas have been observed in distant galaxies (some so brilliant they outshine their entire galaxy for a few weeks). But, in the absence of detailed observations of a supernova in our galaxy, astrophysicists have developed complex computer programs that model every microsecond of the collapse and subsequent explosion. It will be interesting to compare such computer models with observations of a nearby supernova once we see one.

When a massive star collapses, extreme conditions are created. The temperature and pressure become enormous, so great that after the collapse even elements heavier than iron are produced by nuclear transmutations in the shell of the exploding matter. All the metallic elements we value so highly—nickel, silver, gold and uranium—were created in such supernova explosions and were expelled into space, some of them eventually becoming parts of new stars. Astrophysicists calculate the relative abundance of some ninety elements that are made in stars and supernova explosions. Remarkably, these calculated abundances match those observed in nature—thus providing some confirmation that the models are reasonably correct. One particular heavy element, technetium, is radioactive with a half-life of 200,000 years, a half-life short enough to imply that all of it decayed away long ago here on earth. But the spectral lines of technetium can still be observed in red-giant stars—direct evidence that stars create new elements.

Detailed models of what happens after the collapse differ. One model, pioneered by Hans Bethe of Cornell University, implies that the material of the outer part of the star bounces off the collapsed core into space. Sterling Colgate, of the Los Alamos National Laboratory, developed a different model in which a burst of energetic neutrinos—subatomic particles created by the nuclear reactions in the collapsing core—literally blows away the outer layers of the star with a neutrino wind from the core. Conceivably both mechanisms—a bounce and a neutrino wind—are at work blowing off the outer envelope.

But, all models predict that the core, which is the remnant of the supernova, becomes a new state of matter—a neutron star. Objects of this kind were theoretically postulated back in 1933 by the astrophysicists Fritz Zwicky and Walter Baade and, independently, by Lev Landau, a Soviet physicist. These scientists wanted to go beyond Chandrasekhar's work. But what is a neutron star?

In a white dwarf, it is the Fermi pressure of electrons that resists the pressure of gravity. But if gravity is strong enough—as it is in stars that explode in supernovas—the electrons effectively get squeezed into protons (a particle found in the atomic nucleus) and turn the protons into neutrons (yet another nuclear constituent). Neutrons, like electrons, also obey the Pauli exclusion principle—you cannot put two neutrons in the same state one on top of the other. It is the resulting neutron Fermi pressure that resists the force of gravity and stabilizes the neutron star.

Neutron stars are not ordinary objects, and their properties boggle the imagination. The term "star" is a bit of a misnomer, since these objects are not true stars. A cubic inch of the nuclear matter of a neutron star weighs 10 billion tons. They are spheres about a dozen kilometers in diameter—the size of a city. But no one is going to visit a neutron star and walk around it; they are good places to avoid.

Using the laws of elementary particle physics, theoretical physicists led by Malvin Ruderman of Columbia University constructed models of neutron stars. They suggest that there is a kind of iron crust on the surface of the star which is extremely smooth and a yard thick. On the crust there may be tiny "mountains" only a millionth

Emitted radiation

Spin axis

Pion condensate

Crust

Superconducting matter

Magnetic field lines

A spinning neutron star or pulsar, about six miles in diameter, is accompanied by a strong magnetic field that whips around with it. Electrically charged particles following the lines of the magnetic field produce a "lighthouse beam" of radiation that can be detected on earth. The surface of the star may consist of a crust of iron nuclei. Below this is a kind of "crystal" of other atomic nuclei and subnuclear particles. As one proceeds to the interior, the matter becomes superconductive—it conducts electric currents without resistance. The very center of the neutron star may consist of a "pion condensate" of subnuclear particles.

of a centimeter high. Yet to "climb" such a mountain would require the energy output of a large city for a full year because the gravity is so great. Most of the interior of a neutron star below the crust consists of nuclei and other subnuclear particles packed down to nuclear density to form a solid "crystal" of nuclear matter. But physicists suspect that a few kilometers below the surface the matter takes on yet another property—it becomes superconducting, meaning it conducts electricity without any resistance. Huge electrical currents can thus flow without loss in the interior of neutron stars, and such currents produce correspondingly huge magnetic fields, which play an important role in generating the observed pulses emitted by neutron stars.

Some physicists speculate that the center of a neutron star consists of a pion condensate—a new state of matter. Pions are subnuclear particles which have been observed in accelerator laboratories and can be thought of as the glue that holds an atomic nucleus together. Under extreme conditions like those found in the core of a neutron star, pions will condense to form a kind of gas capable of supporting the enormous weight. The extreme conditions inside a neutron star push physicists to the boundaries of their knowledge of subnuclear physics. Some suspect that the very core of a neutron star consists of the quark constituents of nuclear particles. Even while the features of the interior of a neutron star are still being debated, most physicists are excited by the idea that neutron stars provide a kind of "natural laboratory" for testing their new ideas about the subnuclear world.

All this is theory. Do neutron stars really exist? Indeed they do; they were serendipitously discovered in 1967. Here is a piece of the story.

Antony Hewish in Cambridge, England, led a team designing a radio telescope—a 4½-acre field covered with 2,048 antenna rods which was to be used to identify distant quasars by their scintillations. It was his extreme good fortune to have Jocelyn Bell-Burnell, a twenty-four-year-old graduate student, on his team. Examining the output of the antenna which swept the sky as the earth rotated, she observed a "bit of scruff"—a distinctive radio signal—coming from a particular spot in the sky. It would be rather easy to disregard such a signal as nonsense noise. The actual output of the antenna was recorded as a line trace on a paper roll, and the "bit of scruff" was just some short jumps in the trace on hundreds of yards of paper, every inch of which was examined by Bell.

A month later, she saw the signal again and soon thereafter analyzed the "scruff" in detail. She saw that it consisted of periodic pulses about one second long. When she reported this result to Hewish, her thesis adviser, he replied, "Oh, that settles it: it must be manmade." Hewish at first thought the signal was being picked up from a local source like a faulty car ignition. But after ruling out all such possibilities, Hewish checked the timing of Bell's pulsed signal and found to his amazement that it was keeping time to 1 part in 10 million. No signal from a terrestrial source could have that precision. The excitement began.

Hewish's team considered the wild possibility that the pulsed signal was being sent by an extraterrestrial civilization perhaps trying to communicate with other societies. However, Bell soon found another such pulsating source. Shortly thereafter, a total of four were detected in different parts of the sky, which made the idea of an extraterrestrial civilization unlikely. It was clear that a new kind of astronomical object had been discovered. Announcing their discovery in 1967, the Cambridge group suggested that the pulsars, as they were then called, could be neutron stars—the collapsed remnants of exploding stars, theoretically postulated back in 1933 by Fritz Zwicky and Walter Baade. Their guess turned out to be correct. Today, we know of more than three hundred such neutron stars, each the pulsating remnant of a supernova explosion.

What causes the neutron star to pulse so rapidly? F. Pacini, an Italian astrophysicist, and Thomas Gold, an astrophysicist at Cornell University, offered an elementary explanation in 1968. Large stars rotate and have magnetic fields similar in shape to the earth's magnetic field—what is called a "dipole" field. Should a large star collapse to a tiny neutron star, as happens in a supernova, then the spin of the star increases enormously, just as the spin of a figure skater increases as she pulls in her arms and legs. The magnetic field which is bound to the star collapses with it and greatly increases. But the axis of spin and the axis of the magnetic field (determined by the north and south poles of the neutron star) need not coincide. So as the neutron star spins rapidly, the magnetic field, on its separate axis, is whipped eccentrically around with it. Electrically charged particles in the vicinity of the neutron star fall into it, producing a beam of radiation that rotates with the neutron star like a beam of light from a lighthouse. This "lighthouse effect" results in the pulsed radio signal first seen by Bell. The pulsation rate of the radio signal exactly corresponds to the extremely rapid rotation rate of the neutron star.

Most neutron stars rotate a few times in one second. But recently, one was detected that rotates at the incredible rate of 640 times in one second! It's hard to imagine a sphere the size of a city spinning around that fast. This "millisecond pulsar" was probably a member of a binary star system from which, in the course of completely consuming its companion star, it picked up its enormous spin.

While theoretical physicists built models of these bizarre objects, the observational astronomers were also busy. Perhaps the most dramatic confirmation that pulsars were supernova remnants came from the optical observations of Don Taylor, John Cocke and Michael Disney. They studied Baade's star in the center of the Crab nebula, the tenuous remains of a supernova observed and recorded by the Chinese historian Toktagu in 1054, subsequently identified as a pulsar with a frequency of thirty times a second. Taylor, Cocke and Disney decided to have a close look at the visible-light output of the Crab pulsar (rather than the radio output), using a conventional telescope at the University of Arizona's Steward Observatory. They hooked up an electronic synchronization system which effectively blinked the detecting apparatus of the telescope in time with the pulsar's known period. If the blinking occurred synchronously with the pulsar's "on" phase, light would be detected, and if synchronously with the pulsar's "off" phase, no light would be detected. In this way the astronomers hoped to test the idea that the Crab pulsar was actually pulsing out visible light the same way it pulsed out radio signals.

If the star indeed blinked, a pulse would appear along a line of little green dots on the recording scope the astronomers were looking at. On the night of January 15, 1969, they decided (after some preliminary difficulties) to have another go at it, and their assistant had by chance left a tape recorder on, so the moment of actual discovery is recorded. Disney's voice (he is English and the accent is unmistakable) comes on: "We've got a bleeding pulse here."

Cocke responds, "Hey," and after a time, not able to contain himself, "Wow! You don't suppose that's really it, do you? Can't be!"

After more tests and checks to make sure the effect they are seeing is real and not a bug in their electronics, Disney says, "God, just come and look at it down here," and they both laugh. "This is a historic moment!"

"Hmmm," cautions Cocke, "I *hope* it's a historic moment." But within days their discovery that the Crab pulsar blinked its visible-light output was confirmed by other observatories. The bizarre idea that a neutron star is a supernova remnant got its most dramatic support.

Today astronomers have discovered hundreds of pulsars in our galaxy, and there may be millions of them—the remains of once-blazing stars. An idea that had been on the fringe of theoretical physics was drawn into the center of observational astronomy.

Beginning in 1969, astronomers noticed "glitches," a rapid speeding up of the spin of some neutron stars. To some astronomers' surprise, these glitches recurred. What could cause them? To explain them astrophysicists made detailed mathematical models of neutron stars. According to some theories the glitches are due to cracks' forming in the solid crust of a neutron star—neutron-star quakes. These cracks cause the neutron star to shrink ever so slightly, and like that twirling ice skater pulling in her arms, the neutron star spins faster. These neutron-star quakes are triggered by complex instabilities in the bizarre interior of the star and give scientists an opportunity to test their theoretical model of the interior.

One of the most beautiful confirmations of the properties of neutron stars came from the observation of "X-ray bursters." In the late 1970s, scientists designed a number of earth-orbiting satellites with the capability of detecting X rays and found many X-ray sources exhibiting mysterious periodic variations in intensity. Some of these bursters emitted fluctuations in X-ray intensity with a frequency measured in seconds; others dramatically increased their intensity every several hours. What could cause such regular, periodic behavior?

Today the consensus is that the X-ray bursts, whose energy in a few seconds equals that of the sun in two weeks, are due to neutron stars orbiting ordinary stars—an unusual binary star system. The explanation of the short and long periodic bursts is as follows.

If the neutron star is rather young it has a characteristic large magnetic field at its north and south poles. As ionized hydrogen gas is gravitationally pulled off the neutron star's nearby companion, it is drawn by the strong magnetic field to the two poles—a steady stream of hot gas emitting X rays and pulsing with the rotational frequency of the neutron star, about a few seconds.

On the other hand, if the neutron star is old its magnetic field is much weaker. Then something very remarkable happens. The hydrogen gas pulled off the companion star gets distributed over the whole surface of the neutron star rather than concentrating just at the poles. This matter sits on the surface and does not emit X rays until after a few hours a critical amount has accumulated. Then in a spectacular "thermonuclear flash" it explodes all at once, producing an X-ray burst that can be seen here on earth. The process continues to periodically repeat—an impressive confirmation of interconnected theoretical work on neutron stars and observations based on X-ray astronomy.

In view of the sensational results learned from X rays, astronomers have begun to look at light at energies even higher than that of X rays by building gamma-ray spectral-line detectors and sending them aloft in balloons and satellites. Gamma rays, a form of light of very high energy, yield important new information about neutron stars or black holes which have the energetic processes that can emit them. Gamma rays come from atomic nuclei or electron–positron annihilation and hence are independent of the chemical state of the matter. They provide yet another set of detailed "fingerprints" to help identify the complex physical processes that surround exotic compact objects. Still in its infancy, gamma-ray astronomy is growing rapidly.

The idea of neutron stars was born in the imagination of theorists who persisted in asking "What is the final state of matter—what happens when massive stars can support their weight no longer?" In order to answer such questions they turned to examining the growing knowledge of atomic and nuclear physics—knowledge confirmed by detailed laboratory experiments—and in the case of white dwarfs and neutron stars, their answers proved correct. But theoretical physicists knew that for more massive stars even the formation of a neutron star or white dwarf is insufficient to halt collapse. As far as they knew there was nothing that could stop the crush of gravitation. Yet something *had* to happen. What happens, according to the theorists, is that a black hole forms—space and time in a sense collapse to create an object from which not even light can escape.

BLACK HOLES

All light emitted from such a body would be made to return to it, by its own power of gravity.

—John Mitchell, rector of Thornhill in Yorkshire, 1784

. . . that the attractive force of a heavenly body could be so large that light could not flow out of it.

—Marquis de Laplace, 1798

Unlike the case for white dwarfs and neutron stars, there is no universally accepted evidence for black holes, although many astronomers are convinced they must exist and some believe that they have already been found. Black holes are a prediction of Einstein's theory of general relativity—the modern theory of gravity. General relativity has been experimentally tested with brilliant success. Yet some critics would point out that the experimental tests, in spite of their success, are all done for rather weak gravity fields and the theory has never been tested for superstrong gravity fields—the kind one might encounter with black holes. But if the theory indeed applies to such strong fields, then there is no way out of the conclusion that black holes must result when very massive stars collapse.

The proper description of the properties of black holes is given by Einstein's theory of general relativity, which specifies the curvature of space associated with gravitational fields. General relativity is a highly mathematical theory, not easy for an outsider to grasp. But the physical picture of the world it describes—the world of curved geometrical space—can be understood rather simply. How do scientists envisage curved space?

Imagine shooting laser beams through empty space in order to map out the geometry of the space. In ordinary flat space they would go in straight lines so that parallel beams never meet. Suppose that we measure the beams very precisely and find that they do not travel in straight lines but gently curve. We next conclude that this curving of the beams is due to the intrinsic curvature of space in the same way that a jet plane, traveling between two distant cities on the globe, follows a curved path because of the curvature of the surface of the earth. By shooting laser light beams in all directions we can map out the curvature of this three-dimensional space. This is analogous to rolling a small ball along a curved surface. By examining the paths the ball rolls along, we can map out the curving surface. Using light from distant stars or radar from earth (which travels the same way as a light beam), scientists show that actual space does deviate from flatness near large masses like the sun. The central idea of general relativity is that the curvature of space and its influence on the motion of particles or light rays is completely equivalent to gravity. The intense gravity of the sun causes a small but measurable bending in the path of a light ray.

Imagine next an extreme situation in which the whole mass of the sun is crushed down to a radius of a few kilometers. The gravity and space curvature near this compacted sun is enormous. If a light beam were sent out to hit and bounce off this object it would never return. The bending of space becomes so great that even light gets caught by gravity. The actual orbit of a grazing ray of light is a spiral into the object. Since light cannot leave this object, it "appears" as a black hole in space.

The boundary of a black hole, which is not an actual material surface but simply a mathematical boundary within which no light escapes to the outside, is called the "event horizon"; any event occurring inside this boundary can never be observed from outside the boundary. The event horizon is a one-way gate—you can go in but you can never get out.

What's it like to fall into a black hole? If the black hole is small— like one that formed from a collapsing star—then you are in deep trouble. The tidal forces near a black hole are enormous, and the gravitational gradient over even a distance of a few centimeters is enough to tear anything apart.

Time undergoes strange distortions as well. An observer, provided he was not torn apart, who fell into the center of a black hole could see time slow down. But the falling observer can never communicate his strange experience to his friend outside. An outside observer watching his helpless friend fall into the hole would see him take a long time to cross the event horizon.

As black holes get more massive, they also become larger and less dense. If black holes exist which are as massive as trillions of solar masses, then one could fly across the event horizon and suffer no ill effects. But still there is no way out, and only a few minutes would pass before disaster strikes. One would encounter a space-time singularity—a point at which energy density becomes infinite—believed to lie at the very center of the hole. One can even imagine larger black holes. Conceivably, the entire universe is in the process of becoming a giant black hole and we are living inside it—a universe that someday will stop expanding and collapse upon itself.

It wasn't until the 1960s and 1970s that theoretical physicists and astrophysicists appreciated the bizarre properties of black holes, even though those properties could have been directly deduced from Einstein's gravitational theory of 1915. Part of the reason for this delay of half a century was that there was no unambiguous observational evidence for black holes which would force them to think about this possibility. Sometimes physicists just have to have their noses rubbed in a new discovery before they take their own theories seriously. A further reason for delay was that the conclusion that black holes existed required pushing the theory of relativity to extreme limits, and many people were unsure that the theory was applicable in such extreme circumstances. Interestingly, the possibility that black holes really exist was vitalized by the dramatic discovery of neutron stars. If something as bizarre as a neutron star can exist, why not a black hole? Today black holes have gone from the speculative fringe of physical theory to the center. Not a year goes by without theorists evoking black holes to explain yet another unusual astronomical observation.

One of the early pioneers in developing the theory of the black hole was the German astronomer Karl Schwarzschild. In the winter of 1915, he found a simple but exact solution to Einstein's equations of general relativity which described the curvature of space around a spherically symmetric mass like the sun—an accomplishment that impressed Einstein. Schwarzschild found that for a sufficiently compact mass there was a finite radius (now called the Schwarzschild radius—the radius of the event horizon) at which emitted light waves would have an infinitely long wavelength—which is the same as saying that light cannot escape.

The next major step was taken by J. Robert Oppenheimer in 1939 in two papers, the first written with George M. Volkoff and the second with Hartland Snyder, both graduate students. They were motivated not so much by Schwarzschild's work as by a desire to understand what happens when a star collapses. The first paper examined the consequences of the gravitational collapse of a star in a supernova. But Oppenheimer and Volkoff recognized that under special conditions the pressure of gravity even prevents the formation of a neutron star, which would halt the collapse and "the star will continue to contract indefinitely, never reaching equilibrium."

In the second paper Oppenheimer and Snyder went a step further by analyzing the details of the endless collapse. They realized that an inside observer collapsing with the star would never be able to tell an outside observer what his fate was. No radiation escapes from such an object, and "The star," they commented, "thus tends to close itself off from any communication with a distant observer; only its gravitational field persists." While the existence of such bizarre objects was now convincingly shown to be a logical consequence of general relativity theory, no one, Oppenheimer and his collaborators included, went so far as to suggest that they actually existed in the universe. Instead, these objects were relegated to the realm of mathematical curiosities. And there they remained for many decades.

In 1963, Roy Kerr, a New Zealander, found an exact mathematical solution to Einstein's equation which described a rotating black hole. This remarkable result went beyond Schwarzschild's earlier solution, which described only unrotating masses. Kerr's mathematical work implied that it was possible to extract energy from a rotating black hole. As energy was extracted, the rotation slowed. The Kerr solution stimulated much interest in these "mathematical curiosities." As S. Chandrasekhar, who has contributed so much to modern astrophysics, remarked:

> In my entire scientific life, extending over forty-five years, the most shattering experience has been the realization that an exact solution of Einstein's equations of general relativity, discovered by the New Zealand mathematician Roy Kerr, provides the *absolutely exact representation* of untold numbers of massive black holes that populate the universe. This "shuddering before the beautiful," this incredible fact that a discovery motivated by a search after the beautiful in mathematics should find its exact replica in Nature, persuades me to say that beauty is that to which the human mind responds at its deepest and most profound.

John A. Wheeler, that visionary of American theoretical physics, promoted further interest in these bizarre objects and had the intellectual courage to urge scientists to take them seriously. Addressing the National Academy of Sciences in 1973 upon the occasion of Copernicus' five-hundredth birthday, Wheeler used the term "black hole," —a term he had coined in a 1968 article, and a name that has stuck ever since. He told the scientists assembled there that the idea of total gravitational collapse confronted physicists with a major challenge.

To meet that challenge, theorists have since deepened our understanding of black holes. Much of their work was inspired by Stephen Hawking, a brilliant English physicist at Cambridge University. Already famous for mathematical work he had previously done, Hawking turned his talents to investigating black holes. He, and independently Jacob Bekenstein, an Israeli physicist, discovered a surprising relation between black holes and entropy—the thermodynamic measure of the degree of chaos of a physical system. One might wonder, What do black holes, a consequence of the theory of gravity, have to do with entropy, a thermodynamic quality? Yet there is a deep connection.

To elucidate this connection I must say a few things about entropy —a measure of the messiness of physical systems. Highly ordered systems, such as a crystal with its atoms neatly arranged, have low entropy, while highly disordered systems such as gases, with their atoms flying about chaotically, have high entropy. The second law of thermodynamics states that the entropy of a closed physical system never decreases—things can become more chaotic, but never less. A consequence is that information about the detailed structure of a physical system always tends to deteriorate; in fact, loss of such information (properly defined) is precisely proportional to the increase in entropy of a physical system.

Now we can understand how entropy is related to black holes. If something falls into a black hole it is lost forever—there is no way for someone outside the hole to retrieve it. In particular, information must be forever lost as physical objects fall into a black hole, and the loss of such information implies an increase in the black hole's entropy. What Hawking and Bekenstein showed was that the entropy of a black hole was proportional to the surface area of its event horizon—the boundary of the hole. Since according to the second law of thermodynamics entropy always increases or remains the same, it seems that black holes always had to increase their surface area and hence always became larger. There seems to be no way to get rid of a black hole. But this conclusion is not true. Remarkably, a black hole, if left alone, will eventually evaporate into radiation. How can we understand this?

Hawking, pursuing the thermodynamics of black holes, knew that it was possible to assign a temperature to a black hole which was inversely proportional to its radius. Hawking further realized that any object that has a temperature must emit radiation, just as hot coal emits red light. But the whole idea of a black hole was that nothing could escape from it, including radiation. So there seemed to be a puzzle: how could black holes radiate?

Hawking, to the amazement of other scientists, solved this puzzle in 1974 by finding the means by which black holes radiate precisely the amount required by their temperature. His argument can be simplifed as follows: While it is true that any radiation inside the event horizon—the surface of the hole—cannot escape, radiation just outside the hole can. Hawking argued that the intense gravitational field just outside the surface of the hole could spontaneously create a particle and its antiparticle. Similar creation processes are actually required by the quantum-field theories of elementary particles and have been observed in the laboratory. According to Hawking, one particle of the created pair falls into the hole—lost forever—while the other escapes and can annihilate with an escaping antiparticle, converting into pure radiation. The radiation that escapes is now called the "Hawking radiation." For large black holes which might form from collapsed stars, this radiation can be calculated, and its intensity is minuscule. But micro–black holes are "hot" and radiate away their mass quickly in a spectacular burst of Hawking radiation. Micro–black holes that might have been created at the time of the big bang could be exploding only today. The telltale bursts have been looked for but not seen. Perhaps only the very stable large black holes exist today and all the micro-holes evaporated long ago.

With all these fascinating theoretical developments, black holes moved from "mathematical curiosities" to the center of speculative astronomy. Most physicists think they must exist, but no unambiguous observational evidence is yet available to confirm this theoretical prejudice. Of course we will never "see" a black hole because there is nothing to see. How, then, can black holes be detected?

Black holes do have observational signatures, foremost of which is the enormous amount of radiant energy (which is *not* the Hawking radiation) that leaves the neighborhood of a black hole. This radiation arises from matter falling into a black hole. A black hole is like a cosmic vacuum cleaner which sucks up anything in its neighborhood. A black hole in close orbit about an ordinary star thus pulls matter off the star. As accelerating matter falls into the black hole it emits lots of radiation, including X rays, which can be detected here on earth. Today black holes are routinely invoked as possible explanations for the observation of enormous bursts of energy originating from a small volume in space. Nothing short of a black hole seems able to account for such energy release.

There are two places in which black holes, if they exist, ought to be observable. The first is in orbit about an ordinary star, where they consume the matter of the star, thus emitting radiation. The second place is in the core of galaxies, where large black holes may be tearing apart whole stars. Let us examine the evidence for these in turn—evidence that has been provided by an entirely new scientific field, X-ray astronomy, one of the most exciting leaps of astronomy in the last decades.

X rays are a form of light but with a wavelength far shorter than that of visible light. There is every reason to expect that the heavens are filled with X rays just as they are filled with visible light. Unfortunately for astronomers, X rays do not get through the earth's atmosphere, and for many years the existence of cosmic X rays was a matter of conjecture. In the early 1960s, rocket and balloon flights above the atmosphere provided the first glimpse of the X-ray sky. Amazingly, sources of X rays were detected, including the source in the middle of the Crab nebula, which we now know harbors a neutron star. Although these rocket flights provided observations lasting only about five minutes, they already showed that dramatic changes were occurring in the few X-ray sources they had spotted.

Encouraged by these results, a group of scientists from Cambridge, Massachusetts, who had organized a research group known as American Science and Engineering proposed that NASA launch a small satellite in orbit about the earth that could have an extended look at the X-ray sky. Led by Riccardo Giacconi, the group of scientists launched a satellite called Uhuru (Swahili for "freedom") from an offshore drilling platform near the coast of Kenya on December 12, 1970—Kenya's independence day. The cost of this project, including the spacecraft, launch vehicle and payload, was about $13 million. Never in recent times had so much astronomical knowledge been acquired for such a modest expense.

Uhuru, which scanned only a fraction of the night sky, spotted 125 distinct X-ray sources in the first seventy days of observation. The heavens were ablaze with X-ray sources! It was as if a veil covering the sky had lifted. If we could orbit above the atmosphere and be outfitted with special glasses so that we could see X rays, a whole new and unfamiliar heaven would appear. Astrophysicists are still digesting the data sent by Uhuru and have subsequently launched satellites such as the *Einstein* observatory, the *SAS-3* satellite, the Japanese X-ray satellite *Hakucho* and many others. Today more than 300 distinct sources of X rays in the sky are known. They provide a deep clue about the nature of the universe.

Many of these X-ray sources are pulsars, easily identified because of the extreme regularity of their pulses due to their spinning. The location of about a dozen of these X-ray pulsars have been pinpointed to such accuracy that optical astronomers can turn their telescopes to the point and identify the visible counterpart. Sometimes astronomers observe the X-ray and radiowave intensity from such a pulsar increasing by a factor of more than 1,000. Astronomers suspect that this occurs when the "hot spot" on the neutron star (its north or south magnetic pole, where matter most easily falls into the star) is facing the earth, so that we get a direct blast from the beam of X rays and radio waves.

Other X-ray sources do not fit into any neat category. Some of these are binary systems—a pulsar in orbit about an ordinary star. The compact and massive pulsar draws matter off its companion, and this process results in the emission of X rays. The most powerful X-ray source, SCO X-1, is believed to be such a binary system with the pulsar (or perhaps it is a white dwarf) in a tight orbit about an ordinary star. Large amounts of gas are drawn into the pulsar and are responsible for great blasts of X rays. SCO X-1 is so powerful an X-ray source that it can disturb long-range radio communications on earth as it passes over us in the night sky.

From the earliest days of X-ray astronomy it was clear that an X-ray source called Cygnus X-1 (Cyg X-1) was unusual in its observed output. No definite regular pulsing was observed, so it could not be identified as a pulsar. What the X-ray astronomers saw was large fluctuations of intensity lasting anywhere from one-twentieth of a second to ten seconds and small periodic pulse trains which lasted only a few seconds. Because of these odd results, optical astronomers turned their attention to this unusual object and found a visible counterpart—a star catalogued as HDE 226868—which might be at least 30 times the mass of the sun. Most astronomers concluded that the X rays were due to a binary system consisting of a very massive companion in a tight orbit about the star HDE 226868—so tight it might be as little as one-third the distance of Mercury's orbit from our sun, thus distorting the shape of the star and pulling huge amounts of gas off it. Analysis of the signals suggested to many astronomers that the companion was extremely massive and very compact—so much so that it could not be a neutron star; it had to be a black hole.

In November of 1982, observations made at the Cerro Tololo

Interamerican Observatory in Chile suggested a second black-hole candidate called LMC X-3 in the Large Magellanic Cloud. This black hole is in a tight orbit about a normal star, circling it every 1.7 days, swallowing gas and emitting X rays. By measuring the velocity of the visible star the astronomers can determine that its companion has a mass of about 10 solar masses—too large for it to be a neutron star. Eventually the black hole will completely swallow its companion, becoming isolated and completely invisible. There may be many such black holes, which, having completed the feast of their companion, are now dark solitaries wandering around the galaxy.

The radical conclusion that Cyg X-1 and LMC X-3 are black holes is not shared by all astronomers. Some think that a neutron star could do the job of producing the observed X-ray signals. It is fair to say that the data from Cyg X-1 and LMC X-3, while consistent with the black-hole hypothesis, do not prove that they are black holes. It is likely that the question of the existence of black holes will be open for quite some time.

What could settle the issue? Theoretical astrophysicists are building mathematical models of black holes surrounded by gas in an attempt to find the distinctive "signatures" of black holes—signals that could be attributed only to black holes. Such signatures could be signals from the hot gas as it gets sucked up into a black hole. According to one model, the gas around a black hole should form an "accretion disk" similar to the disk of rings around the planet Saturn, and hot spots—intense sources of X rays—might appear there. These hot spots would orbit the hole with the disk in one-thousandth of a second and ought to emit rhythmic pulses at that rate before they are swallowed by the hole. Such accretion disks can also form around pulsars, and the properties of such disks are similar to those surrounding black holes. This makes it hard to distinguish the two.

To look for such rhythmic pulses from accretion disks was the task of *HEAO-1* (High-Energy Astronomy Observatory), an X-ray satellite launched in 1977. But no such millisecond rhythmic pulses coming from Cyg X-1 or any other black-hole candidates in binary systems were seen. However, other X-ray sources called bursters did exhibit rhythmic X-ray trains which some scientists thought might be the signal of black holes. But this explanation of the X-ray bursts is not widely held now. Instead, they are now understood to be due to neutron stars in orbit about ordinary stars. The evidence for black holes in binary systems remains uncertain.

A black hole in orbit about a companion star drawing matter off of it. The captured matter—hot gas—forms an accretion disk around the hole before being swallowed by the hole. According to some theoretical models, radiant energy is ejected by the black hole's accretion-disk system along its spin axis.

The black holes that some astronomers conjecture to lie in some binary systems are small fry compared with those monsters they believe to lie in the nuclei of galaxies—supermassive black holes millions of times the mass of the sun. Astronomers know there is something going on in the cores of galaxies which releases enormous energy, and it could be gigantic black holes. Those who don't accept the reality of black holes simply refer to the object or process in more neutral terms such as "the machine," "the monster" or "the prime mover." But no one doubts that there is something bizarre to be found at the center of galaxies.

The nearest galactic nucleus is, of course, the nucleus of our own galaxy—the Milky Way. But we cannot see the nucleus directly because it is obscured by dust and gas. Yet radio waves and infrared and X rays which penetrate the dust and gas can tell us about the core. We now know that there is a massive concentration of stars in the galactic core. If our solar system were located there, our night sky would be lit with the brightness of a hundred full moons coming from all the stars in the core. A number of infrared sources have been detected, and these are probably enormous red-giant stars. Observations of ionized gas suggest the existence of a single ionizing source which can be identified with a strong compact radio object—presumably a black hole—at the very center of our galaxy, a place known as "Sagittarius A West." According to the black-hole model, a central supermassive black hole is heating the surrounding envelope of gas, and this gas is the source of the X rays and ionizing radiation. The black hole itself is rather small—the size of the solar system—and is rather well behaved. But it is surrounded by violence.

In December of 1983, Cal Tech astronomers using the twenty-seven radio antennas of the Very Large Array in Socorro, New Mexico, scanned the Sagittarius A West region and got a high-resolution image. They found that three arms of rapidly moving gas are swirling (and presumably falling) into the point believed to locate the black hole. If this interpretation is correct, then as one of the Cal Tech astronomers said, this is "the first time we are actually seeing matter falling into a black hole." In 1984, astronomers went on to find that huge magnetic fields surrounded the core—a finding that has them puzzled.

Other evidence for black holes in galaxies comes not from our own galaxy but from the investigation of extremely active, presumably young galaxies. (The Milky Way is a rather mature galaxy). There is

a growing opinion that quasars—enormously distant, highly energetic astronomical objects—are an early stage in the development of
galaxies. Could quasars be a manifestation of the dynamics of black
holes? Little else seems capable of explaining the prodigious energy
output of these remarkable objects. Quasar black holes could be
fueled by infalling gas, some of which is subsequently thrown back
into space and forms a "cosmic blowtorch"—a jetlike structure
which resembles the celestial objects that are observed. All the active
galaxies, the violent Seyferts, the BL Lacertae objects, the quasars—
all these may be just different evolutionary stages in the development
of galaxies—stages in which "the machine" in their core has not yet
quieted down to middle age as it has in our galaxy.

The role of supermassive black holes in active galaxies is, of course,
speculation. But if astronomers had to place bets on the best place to
find such a centrally located supermassive black hole, most would
bet on the core of the galaxy M 87—a giant elliptical galaxy which
plays a major role in gravitationally binding together the Virgo "supergalaxy" consisting of more than 1,000 galaxies. The core of M 87
is quite active, as revealed by radio and X-ray emissions, and it may
be an old quasar.

Interest in M 87 motivated a 1978 optical examination of the core
by means of the most recent electronic technology that could amplify
light signals. The astronomers found a sharp peak in the light intensity at the core of M 87. They also found that the stars moving near
the center of the galaxy showed an abrupt increase in their speed as
one looked closer to the center—as if they were attracted to a very
massive object in their midst. They concluded that these observations
are entirely consistent with the existence of a supermassive black hole
—5 billion solar masses—at the center of the galaxy M 87, although,
once again, the existence of a black hole could not be definitely
proved. With the advent of the Space Telescope, to be launched in
1986, M 87 will be resolved 20 times better and perhaps we will learn
more.

Recently a group of European astronomers made an extensive and
detailed analysis of data obtained from the *International Ultraviolet
Explorer* satellite launched in 1978 and concluded that one of the
brightest of the energetic Seyfert galaxies, NGC 4151, contained a
black hole of about 100 million solar masses. Through a stroke of
good luck, the satellite was observing this galaxy when its core suddenly flared up. Orbiting clouds of gas surrounding the core were

also sequentially activated, and these data enabled the investigators to determine the velocity and distance of the clouds from the center. Once this information is known, it is an exercise in Newtonian mechanics to calculate the mass of the central object; and this is how the presence of a black hole was established.

Black holes have come a long way from being mathematical curiosities at the fringes of the scientitic imagination. They are now routinely invoked by theorists to explain almost any new observation in astronomy that requires a huge energy source in a small region of space. Yet we do not know for certain if they exist. As Princeton astrophysicist Edwin Turner remarks, "Black holes are a central idea in the theory of active galaxies, and for the empirically minded, direct observation is essential." Wouldn't it be wonderful if a definite proof of the existence of a black hole could be found? Perhaps some incontrovertible evidence is just around the corner. But perhaps not. It is possible that no definite evidence will be found but that scientific opinion will gradually shift in favor of or against the idea of black holes. In a way, that would be a pity. Ideas as dramatic and far-reaching as those which imply the existence of black holes deserve either dramatic and far-reaching observational confirmation or a conclusive rejection. One would like the story of black holes to come to a conclusion with a bang, not a whimper.

Looking back over this century we can see how far astrophysicists have advanced our understanding of stars—how they are created and how they live and die. The major features of stellar evolution are now known. And not only do the stars seem less mysterious now, they also seem friendlier. Like us they are born, live and die, and like us they survive in a larger society—the galaxy.

The study of galaxies is a branch of astronomy which unlike astrophysics is still in its infancy. No systematic understanding of the evolution of galaxies—one that would take us from the primordial seeds from which they grew in the big bang to their present magnificence—yet exists. That is for the future. But much has been learned about galaxies, most of it in the last decades, and it is to this topic that we now turn.

4

The Discovery of Galaxies

For I can end as I began. From our home on earth we look out into the distances and strive to imagine the sort of world into which we are born. Today we have reached far out into space. Our immediate neighborhood we know intimately. But with increasing distance our knowledge fades . . . until at the last dim horizon we search among ghostly errors of observations for landmarks that are scarcely more substantial. The search will continue. The urge is older than history. It is not satisfied and it will not be suppressed.

—Edwin Hubble, from his last scientific paper

The fact that the night sky appears filled with stars supports the illusion that the immense space of the universe must also be uniformly filled with stars. So persuasive is this illusion that not until this century could astronomers definitely prove that stars are parts of galaxies—the "island universes"—and that galaxies are the principal inhabitants of the cosmos.

If we could step outside the Milky Way, we could see that it is an immense spiral disk, its diffuse arms twisting around a central bulge of stars within which hides the galaxy's mysterious nucleus. The boundary of the bulge is marked by a ring of thick, lumpy clouds of molecular hydrogen. If we look closely, we see that the arms are delineated by bright blue stars and contain lots of dust and gas concentrating in star-forming nebulae. Our own sun is located on the inner edge of one such arm, the Orion arm: one star out of the hundreds of billions that make up the Galaxy.

Although the spiral disk and its central bulge are the most conspicuous features of our galaxy, if we look at the space surrounding the galaxy—the halo—we see that it contains "globular clusters" of stars, each an independent system consisting of 50,000 to a million stars gravitationally bound together to form a roughly spherical cluster. Some astrophysicists think the centers of the globular clusters might contain small black holes of 100 solar masses. Astronomers discovered several hundred of such star clusters (about a hundred in the halo and another hundred in the disk) which arrange themselves symmetrically about the galaxy and tend to concentrate near its center. Each globular cluster is, in effect, a satellite of the Milky Way galaxy. Astrophysicists learned that most of the stars in the globular clusters are very old stars, dating back to the time of the formation of the galaxy itself. The globular clusters are probably a reminder of things past and are a clue, as yet not understood, about the origin of the galaxy itself.

The mistress of this vast galaxy—the disk, the central bulge and its halo of globular clusters—is gravity, the only force guiding the motion of widely separated stars. The law of gravity discovered by Newton is rather simple: the force between two masses is always attractive, proportional to the product of their masses and the inverse square of the distance between them. Yet in spite of its simplicity, the law of gravity must account for the complex configurations of billions of stars following tortuous trajectories and the fact that the galaxy has maintained its shape for billions of years and not collapsed or flown apart. Mathematical physicists and astrophysicists who tackle this problem of galactic structure are aware that trying to track the motion of even a few hundred stars according to the law of gravity taxes the limits of our largest computers. To deal with the larger number of stars appropriate for galaxies, statistical methods, which average over lots of stars, must be used. So in spite of the simplicity of the law of gravity, physicists still find it difficult to account mathematically for such features of the galaxy as its shape and stability. But they do know that the attractive and long-range nature of gravity guarantees that a dynamical system of stars cannot be absolutely stable for all time. A galaxy must change and evolve, perhaps even dramatically.

The standard view of the galaxy held by astronomers for decades

Plan of the Milky Way Galaxy

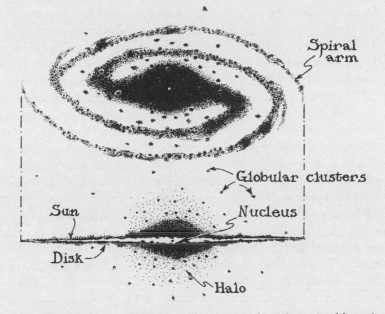

A rough illustration of our Milky Way galaxy as seen from the top and from the side. The diameter is about 100,000 light-years. This shows the main visible components of the galaxy—the central bulge of stars surrounding the nucleus, the spiral arms and the halo of globular clusters, little Independent star systems, concentrating near the center of the galaxy. This illustration, while it reveals the main visible components, does not show the corona of hot gas surrounding the galaxy, the magnetic fields in the disk or the dark matter that extends out far beyond the visible edge of the galaxy. The dynamics of galaxies are just beginning to be understood. The very center of the galaxy may harbor a large black hole.

—that it consists of a central bulge, a disk spiral and a halo of globular clusters—began to change dramatically in the early 1970s. Mathematical studies by Donald Lynden-Bell of Cambridge University and by a Princeton University group which included Jeremiah P. Ostriker, P. J. E. Peebles and Amos Yahil showed that the disk of a spiral galaxy would not be dynamically stable unless the galaxy was surrounded by an extended massive halo of dark matter. If this idea is correct, then nearly all the mass of a galaxy—as much as 90 percent

—lies not in the visible stars and gas but in a new component, the invisible halo.

Observational, rather than theoretical, evidence for the existence of a massive invisible halo came from J. Einasto and his associates at Tartu Observatory in Estonia. Einasto studied the motion of our galaxy relative to nearby galaxies and found that it was moving rather rapidly. He concluded that it had to be far more massive if it was to be gravitationally bound to the system of nearby galaxies. On this basis he suggested that a massive halo supplied the missing mass.

The most dramatic evidence for the existence of the invisible halo came when astronomers measured the velocity of gas that orbits galaxies far from their visible edges. If all the mass of a galaxy were concentrated in the visible stars, then the velocity of the orbiting gas should decrease the farther away it was from the galaxy, just as the velocity of a planet in orbit about the sun decreases the farther away from the sun it is. Instead, the astronomers Vera C. Rubin, W. Kent Ford, Jr., and Norbert Thonnard of the Carnegie Institution in Washington, D.C., were amazed to find that the velocity of the orbiting gas does not decrease but remains constant, indicating that the major mass of a galaxy does not stop at its visible edge but instead extends far out beyond it, an effect already recognized by Martin Schwarzschild, Leon Mestel and others in the 1950s. Most astronomers are now convinced that galaxies have massive invisible halos and that the distribution of light in a galaxy is no indication of the distribution of mass. The most popular recent candidates for the dark matter in the halo are new quantum particles conjectured by theoretical physicists.

When new ultraviolet-light detectors were placed in orbiting satellites such as the *International Ultraviolet Explorer*, astronomers in the late 1970s confirmed the 1956 conjecture of Princeton physicist Lyman Spitzer, Jr., that our galaxy is surrounded by a corona of hot gas extending above and below the disk. This corona, which absorbs ultraviolet light from distant stars and can therefore be detected, is unrelated to the hypothetical halo of invisible matter. Evidently the disk of the galaxy, where all the stars lie, explosively spills out hot gas into space above and below the disk in gigantic streams. Once in space the hot gas cools, loses velocity and falls back into the galactic disk—a cycle dubbed the "galactic fountain." The power for the galactic fountain seems to be supernova explosions of stars in the disk. Such coronas of hot gas also appear surrounding other galaxies.

Not only did the view of the large-scale "architecture" of our galaxy change in the last decade, but astronomers studying the detailed interior structure have also altered their views. The most massive single inhabitants of our galaxies are not stars but the giant molecular cloud complexes that concentrate in the spiral arms. Their existence was heretofore unknown. These cloud complexes are places of star formation and complicated physical processes which play a major role in the evolution of our galaxy. New observations and studies of the nucleus of our galaxy also suggest that it harbors a very massive compact object, perhaps a black hole. As these unanticipated discoveries indicate, scientists are just beginning to come to grips with what is really going on in a galaxy. Much remains to be learned.

The disk of the Milky Way Galaxy is about 100,000 light-years across. If we fly about a million light-years away from our galaxy and look back, we see that it is not alone—it is surrounded by a group of lesser satellite galaxies. Among the most conspicuous of these are the Large and Small Magellanic Clouds (visible from the earth's southern hemisphere), irregularly shaped galaxies near the boundary of the Milky Way which have been torn apart and distorted by gravitational tidal interaction with our galaxy. Still farther from our galaxy are the satellite dwarf galaxies—Draco, Sculptor, Sextans, Pegasus, Fornax, Ursa Major and Minor, Carina, Leo I and II—each a small, sparse galactic system of stars compared with the Milky Way. Probably other such small galaxies exist in our vicinity but have been overlooked because they are too dim to be detected. All these small satellite galaxies are within a radius of 1 million light-years from the center of the Milky Way.

A little more than 2 million light-years away lies another great spiral galaxy, comparable to the Milky Way—our sister, the Andromeda galaxy. Looking at the Andromeda galaxy from earth, we get a good idea of what our own galaxy must look like from afar, because the two galaxies are so similar. The Andromeda galaxy, like ours, is surrounded by lesser satellite galaxies, of which the most prominent are two companion elliptical galaxies orbiting about it. The whole Andromeda system of galaxies is moving toward the Milky Way system at about 55 miles a second, and billions of years from now, the two galaxies may collide. Such a collision will not be the catastrophe one imagines, because galaxies are mostly empty space. The two galaxies will pass right through each other. But the effect on the interstellar gas will be dramatic, and the mutual gravitational interaction between the stars of the two galaxies will distort each galaxy's shape, indicating that a collision indeed took place.

The Milky Way and Andromeda galaxies, along with their associated satellites and a few strays, about seventeen or eighteen galaxies in all, collectively constitute what astronomers call "the Local Group." This is our "home base" in the field of galaxies that populates the entire universe—our corner of the cosmos. If astronomers want to do detailed observations of galaxies they examine the Local Group because these are the nearest galaxies.

Examining larger regions of space beyond the Local Group, we find more and more galaxies; they seem endless in number. But we notice that the galaxies are not scattered at random in space but, instead, that they tend to cluster in groups consisting of several

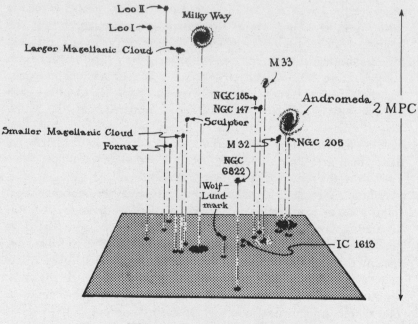

The Local Group

The Local Group of Galaxies. A three-dimensional map of our region of the universe showing our Milky Way galaxy and its neighbors. There are some 100 billion galaxies in the observable universe.

hundred large galaxies accompanied by perhaps thousands of smaller ones. The nearest such cluster of galaxies, the central part of the Virgo cluster, is between 30 and 60 million light-years away and consists of hundreds of large spiral galaxies. Between 200 and 400 million light-years farther away is the Coma cluster, harboring at least 1,300 major galaxies. The universe is populated with such clusters of galaxies. But clusters of galaxies are not even the largest aggregates: the hierarchy of clustering goes on to form superclusters.

Clusters of galaxies such as the Coma and Virgo clusters and many lesser clusters tend to congregate into such superclusters—gigantic clusters of the clusters. Our Local Group of galaxies, for example, is part of such a supercluster, which also includes the Virgo cluster. Such superclusters of galaxies are the largest well-defined objects that bear a human's name—a fact recorded in *The Guinness Book of World Records:*

> Eponymous Record. The largest object to which a human name is attached is a super cluster of galaxies known as Abell 7, after the astronomer Dr. George O. Abell of the University of California. The group of clusters has an estimated linear dimension of 300,000,000 light-years and was announced in 1961.

Galaxies are the primary, visible inhabitants of the universe. These great islands of stars are arranged in a hierarchy consisting of galaxies, clusters of galaxies, and superclusters. Why does the universe arrange itself in this peculiar way? Why, for example, aren't the stars, or even the galaxies, uniformly distributed in space? Some astronomers speculate that the hierarchy of clusterings represents an evolutionary development in the universe. All the structures in the universe are unstable; they change and evolve, albeit very slowly by human standards. According to this view, primordial galaxies formed shortly after the origin of the universe itself, and went through a series of evolutionary stages. Galaxies today have matured, and the era of dramatic galactic evolution is over. We are now in a new era as galaxies within clusters move closer together, perhaps forming into tighter clusters and superclusters.

A still deeper set of questions than that posed by the peculiar distribution of galaxies in space is Why do galaxies exist at all, what is their origin and what are the details of their subsequent evolution? Little is known about the answers to such questions. Walter Baade, one of the great astronomers of this century, remarked that our present understanding of galaxies is as incomplete as our understanding of the stars was at the beginning of this century, before we comprehended the energy release from nuclear burning. We do not yet know the fundamental energy mechanisms responsible for galactic evolution. Yet in spite of our ignorance about their dynamics, it is remarkable how far our knowledge of galaxies has advanced in this century.

The ongoing achievements of modern astronomy are due to two remarkable developments. First is the advent of new instruments—high-resolution telescopes, optical technology, timing mechanisms and more recently, radio telescopes, computers and earth-satellite observatories. The second development is the new knowledge we have acquired about the properties of matter here on earth.

Astronomy is an observational, not an experimental, science. It is not possible to experimentally alter astronomical objects to see how they physically change—stars are too far away and too large for us to do that. But it is possible to do experiments here on earth and learn about the properties of matter—the behavior of atoms, molecules, light and gases. Since these properties are universal, even the matter in the distant stars and galaxies must obey the laws of nature discovered here on earth, and the observations of the astronomers can be interpreted by astrophysicists in terms of familiar physical processes. Terrestrial experiments, supplemented by theoretical computer modeling, go hand in hand with astronomical observation—a powerful combination that opens the cosmos to rational investigation.

The story of the discovery of the external galaxies and the distribution of stars in our own Milky Way is part of the great scientific history of this century. It is a story that is far from over. We are like children who, having learned the environment of their own home and backyard, have seen that there are other, similar houses beyond the fence, and that there is an even wider world, as yet unknown, that lies beyond the neighborhood. Someday we may even learn how that environment was created. To begin this story, let us go back a few centuries to those first philosophers and astronomers whose speculations and observations pointed the way to the modern view of the galaxies.

Thomas Wright, an Englishman from Durham, wrote down his cosmological thoughts in *An Original Theory: or, New Hypothesis of the Universe,* which was published in 1750. From our modern viewpoint it represented a mixture of theology and astronomy; Wright saw the universe as a divine revelation. He supposed that the Milky Way appeared as a band in the sky because it is a flatter layer of stars; one of his models represented it as a "grind stone." In other models he arranged the stars concentrically about a "supernatural" center. Wright speculated that our Milky Way was but one of many such star systems in the universe—an idea that anticipated twentieth-century observations. Later in his life Wright rejected many of his earlier published speculations, but not before a newspaper account of his work, misleading as it was, stimulated the interest of the young Immanuel Kant in far-off Königsberg and set him to thinking about cosmology.

In 1755, when he was thirty-one, Kant published his *Universal Natural History and Theory of the Heavens*, in which he attempted to develop a cosmology that took into account the new Newtonian physics: the motion of the stars in the Milky Way had to be consistent with Newton's law of gravitation. Kant also emphasized that the different kinds of nebulae then observed might require different explanations—something we know to be true today. Some of these, he guessed, were great external star systems. Kant wrote:

> It is far more natural and intelligible to regard [the nebulae] as being not enormous single stars, but systems of many stars, whose distance presents them in such a narrow space that their light, which individually is imperceptible, reaches us, on account of their immense number, as a uniform pale glimmer. . . . And this is in perfect harmony with the view that these elliptical figures are just universes or . . . Milky Ways.

Unfortunately for the young Kant, not only did he publish this important work anonymously, but its publishers went bankrupt and the books got locked in a warehouse. Kant's cosmological ideas did not gain wide circulation until his later fame as a philosopher. In 1761, J. H. Lambert, an astronomer, published ideas at which he had arrived independently but which were similar to Kant's. Kant felt called upon to defend his priority. Again in 1791, William Herschel's observations of the rotation of the inner ring of Saturn—observations that supported some of Kant's ideas—prompted Kant to defend his priority. But most of his specific ideas, especially the suggestion that some nebulae were external star systems, could not be proved until the twentieth century, and in the meantime the nature of the nebulae remained a subject of speculation.

William Herschel, the father of modern observational astronomy, devoted most of the latter part of his life to determining the shape of our star system. He was the first to show that the stars are not symmetrically arranged around the sun. He also concluded that the galaxy has a bilobed structure, a gap in the star field which we know today is due to a dark band of dust-and-gas clouds that stretches from the constellation Cygnus to the southern latitudes. Herschel, like Newton before him, estimated distances to the stars on the assumption they were all as bright as the sun. This assumption is not correct, as Herschel himself came to realize. He eventually despaired of ever determining the precise structure of our star system, the Milky Way.

During the nineteenth century, little progress was made toward understanding of the structure of the Milky Way. Not surprisingly, it is extremely difficult to determine the structure of something if you are *inside* it, instead of on the outside looking at it. In the nineteenth century, it wasn't even clear that there *was* an "outside" to our star system. Most astronomers, until the first decades of this century, thought the universe was a single system of stars which filled infinite space.

In spite of their many misconceptions about the large-scale structure of the universe, astronomers in the eighteenth and nineteenth centuries made many important observations. Among them were the first direct measurements of the distances to nearby stars, which were ultimately important in determining the shape of our galaxy. How can one measure the distance to a far away object like a star without going there? Astronomers used the method of parallax, which is easily illustrated if you hold up a finger about a foot in front of your eyes. Look at the finger with just one eye and then with just the other eye. The finger jumps in position against the background. Knowing the angle subtended by the jump and the "baseline" distance between your eyes, you can calculate by geometry the distance from your eyes to the finger. Astronomers using this method of parallax to measure the distance to stars do essentially the same thing. Employing as a baseline the diameter of the earth's orbit about the sun, they make two observations of the same star six months apart. The angle they measure is the apparent "jump" of the star's position against a fixed background of yet more distant stars. The observation of the parallax would also definitely confirm the Copernican solar model—the earth moves about the sun rather than the sun about the earth.

By the late 1830s, new telescopes had improved sufficiently for the parallax to be seen. In 1838–1839, Friedrich Wilhelm Bessel measured the distance to 61 Cygni, Friedrich Georg Wilhelm Struve the distance to Alpha Lyrae in the constellation Vega, and Thomas Henderson the distance to Alpha Centauri. All these distances were determined to be at least several light-years—even the nearest stars were very far away. It became clear from other measurements that these and other stars in the neighborhood of the sun were moving tens and even hundreds of kilometers per second—and that the sun was part of a dynamical star system.

Perhaps the most intriguing observations were those of the nebulae —a new class of astronomical objects distinct from stars, planets and comets, whose significance remained obscure until this century. One hundred three of these nebulae, fuzzy patches among the stars, were catalogued by the French astronomer Charles Messier in 1784 and his colleague Pierre Méchain added six more in 1786. Messier was a comet hunter (comets were an eighteenth-century obsession with astronomers), and he compiled his famous catalogue ". . . so that astronomers would not confuse these same nebulae with comets just beginning to shine." Subsequently William Herschel published a catalogue of thousands of nebulae, work that was continued by his son, John, who published the General Catalogue in 1864, with 5,000 nebulae and star clusters listed. John Herschel spent several years at Cape Town observing the southern skies, especially the Magellanic Clouds, of which he remarked, "the nebulae are to be regarded as systems sui generis, and which have no analogues in our [northern] hemisphere." We now regard the Magellanic Clouds as independent galaxies.

Expanding on the work of the Herschels, J. L. E. Dreyer published his New General Catalogue in 1890; and the classification system of these works is still used today. For example, the Andromeda galaxy is classified as M 31 (M for Messier), and one of its satellites is NGC 205 (NGC = New General Catalogue). Some of the nebulae classified by Messier and Dreyer are now known to be galaxies far from our own. But not all the nebulae are external galaxies, and the fact that there were very different kinds of nebulae caused a lot of confusion. It turned out that many of the nebulae classified by the early astronomers are actually part of our own galaxy—the "hot spots," the regions in the spiral arms of our galaxy in which gas and dust have accumulated to form new stars like the Orion and Trifid nebulae. A completely different group of nebulae are the "planetary nebulae"— gaseous envelopes of old stars that have been blown into space but looked like the disks of planets (hence the misnomer), or objects like the Crab nebula, which are actually supernova remnants drifting into space that will soon disperse in the interstellar medium. Other fuzzy astronomical objects originally classified as nebulae turned out to be the globular star clusters distributed in our galactic halo. With all these rather different astronomical objects appearing as nebulae, it is no wonder that until recently there was so much confusion as to what they were.

In the 1840s William Parsons, third Earl of Rosse, quit Parliament to pursue his passion for astronomy and used his wealth to build an immense reflecting telescope. This "Leviathan" exceeded the technological capabilities of his time, and while it worked, it was not as useful as smaller telescopes. Yet Lord Rosse, training his monster on the heavens, saw the nebulae in great detail and became the first to see that many of them had a spiral structure. His scientific notebooks are filled with drawings of the spirals. But there was no reason for him to believe that they lay outside our own star system. In fact, there was no reason, until this century for anyone to believe that our star system had an "outside."

After the discoveries of Rosse, photography became increasingly important in explorations of the nebulae. James Edward Keeler, an American at the Lick Observatory in California, continued the nebular photography of the English astronomer Isaac Roberts, who first showed from photographs which resolved the edges of the Andromeda galaxy that it was similar to other spiral nebulae. Keeler photographed more than 100,000 galaxies, most of them spirals. The universe was richly populated with these strange and wonderful objects.

The beginning of this century marked a turning point for astronomical technology. The largest practical refracting telescopes (which use large lenses as the focusing element) were built at the Lick and Yerkes Observatories in the late nineteenth century. If optical astronomy was to progress, then new telescopes had to be reflectors with an immense single mirror as the primary light-gathering and focusing element. The technology for creating such mirrors was now coming into being, and the United States, which was to dominate observational astronomy for the next half-century, began to take the lead over Europe.

A single remarkable individual, George Ellery Hale, an accomplished research scientist from M.I.T., became the driving force of the new American astronomy. Even as a young man, apparently free of those contradictions of purpose which harass most youths, Hale impressed his superiors with his maturity. He solicited the money for the new telescopes from wealthy Americans; he selected the sites for the telescopes, supervised their construction and encouraged the best observers to use them. Hale also promoted the integrity of astrophysics as a scientific discipline in its own right, distinct from astronomy and physics.

Under Hale's leadership, telescopes came into existence that were to reveal the true significance of the nebulae. Why was it so difficult to understand that the spiral nebulae were external to our star system? First, they were extremely distant, so far away that they could not be seen as a system of individual stars (rather than gas) until the deployment of the 100-inch Mount Wilson Hooker reflecting telescope built by Hale in 1919. Second, the spiral nebulae were not uniformly distributed in the sky. There was a "zone of avoidance," a part of the heavens in which the spirals did not appear. We now know that this zone corresponds to the disk of our galaxy and that the dust in the disk obscures any distant spirals. But at the time, the fact that no spirals were seen in the Milky Way region seemed to many astronomers an indication that the spiral nebulae were somehow correlated to the structure of our own star system and not external to it.

These puzzles about the spiral nebulae and the shape of our own star system were not resolved until the 1920s. Perhaps more than any other individual, the American astronomer Harlow Shapley determined the correct structure of our galaxy: a disk of stars with our sun near the edge, surrounded by the halo of globular clusters. How did Shapley come to his remarkable picture of the galaxy which, at the time he proposed it, seemed so peculiar?

Shapley's clue to solving the puzzle of the shape of our galaxy lay in the distribution of the globular clusters—those balls of stars which are distributed in the halo. Prior to Shapley's 1918 work, astronomers knew that the distribution of these star clusters is not symmetrical— most of them are in one hemisphere of the heavens. About one-third of the globular clusters are found in the constellation Sagittarius, which occupies only 2 percent of the sky. Assuming that the center of our galaxy coincided with the center of the distribution of globular clusters, Shapley, using the new Mount Wilson telescope, showed that the center of our galaxy lies in the direction of Sagittarius. By using Leavitt's period–luminosity relation for Cepheids in the globular clusters, he estimated the distance to the center of the galaxy to be very far away indeed. Shapley's view of the geometry of the galaxy subsequently received further confirmation from the Dutch astronomer Jan Oort, who demonstrated the motion of stars about the galactic center.

If Shapley is right, then we are on the fringe of our galaxy. But then other questions arise. Why doesn't the center of the galaxy, with its concentration of stars, glow brightly in the constellation Sagittarius? Why are the stars we see in the night sky distributed so uniformly? The answer to these questions is that our galaxy—like all spiral galaxies—contains lots of gas and dust which obscure the bright starlight from the center of the galaxy. In fact, the gas and dust effectively block out the light from all stars except those several thousand in the neighborhood of our sun which can be seen with the unaided eye, and these appear uniformly in the night sky.

For a long time the dust in our galaxy created a lot of observational complications. Not only does the galactic dust obscure distant stars: it also reddens their light. This reddening made it difficult for astronomers to identify the correct color of a star, a datum that was useful in estimating its distance from us. Only after the careful observational work of Robert Trümpler, a Swiss astronomer at the Lick Observatory, were the complexity associated with the dust and gas and the problem of determining the geometry of our galaxy finally settled in 1930. By studying groups of stars in open clusters, Trümpler demonstrated that in our galaxy starlight has its intensity absorbed by half for every 3,000 light-years traveled in the disk. Using this and other facts, astronomers could "correct" their observations of distant stars and establish accurate distance scales.

Although Shapley was right about the structure of our Milky Way galaxy and defended it vigorously, he ironically persisted in maintaining that the spiral nebulae were part of our galaxy, not external to it. Walter Baade recalled arguing against Shapley's view in 1920, citing the evidence provided by a long photographic exposure taken with a 60-inch telescope which apparently resolved individual stars in the spiral M 33. If true, this meant that M 33 was far away and not part of our galaxy. Shapley thought the images in the photograph were not stars but gas. The differences between Shapley and some of his colleagues came to a head in a famous debate he had with the astronomer Heber Doust Curtis before the National Academy of Sciences in 1922.

Shapley's argument depended on some very precise measurements by the astronomer Adrian van Maanen of the rotation of spiral nebulae. If the spirals were rotating as fast as Van Maanen's measurements indicated, and if one assumed they were very far away (as Shapley's opponent maintained), then the stars in the spiral would have to be moving at velocities comparable to the speed of light, if not faster. This was impossible, because nothing moves faster than light. Hence, either the spirals were nearby objects, as Shapley maintained, or Van Maanen's measurements were seriously off. In spite of the extraordinary care Van Maanen took, his measurements of the rotation of galaxies were at the threshold of the technology of his time and turned out to be wrong. Spiral galaxies do rotate, but very much more slowly than Van Maanen estimated. So Shapley was also wrong, although he did not know it until much later.

Curtis, his opponent in the debate, defended the other view that the spirals were far away and represented separate galaxies, the view we hold today. He focused his argument on M 31, the Andromeda nebula, and argued that it could not be a local gas cloud in our galaxy. But Curtis' evidence was actually rather weak. He pointed out that there were considerable numbers of novae—presumably exploding stars—in the Andromeda nebula, although their observed luminosities, unlike those of ordinary novae, were quite variable. The evidence that these were true star novae like those observed in our own galaxy was not, according to Shapley, compelling, and therefore there was no reason to believe that the Andromeda nebula was anything other than a local gas cloud.

The debate concerning the extragalactic nature of the spiral nebulae was not resolved at the National Academy of Sciences. It was resolved through the careful observational work of Edwin Hubble and many others who devoted themselves to the study of galaxies. Hubble followed up on the lead by the astronomer John C. Duncan, who, while searching for novae in the Andromeda galaxy, had spotted a variable, blinking star which was probably a Cepheid. Hubble found forty further Cepheids in Andromeda and many more in other galaxies. By measuring their rate of blinking, Hubble established their absolute luminosity and thus, by comparing this with their observed luminosity, he could determine the distance to them—a distance so great that the galaxies were certainly very far outside the Milky Way system. According to the distinguished astronomer Allan Sandage, Hubble's observations "proved beyond question that nebulae were external galaxies of dimensions comparable to our own. It opened the last frontier of astronomy, and gave, for the first time, the correct conceptual value for the universe. Galaxies are the units of matter that define the granular structure of the universe." After Hubble's work, other astronomers quickly became convinced that these nebulae were gigantic faraway island star systems similar to our own.

Edwin Hubble lived in the realm of the galaxies. He began, like many other great observational astronomers, in a completely unrelated discipline: he was a student of Roman and English law at Oxford University. After a year of practice he decided to "chuck the law for astronomy." Eventually, he came to use the 100-inch Hooker reflecting telescope at Mount Wilson in California—an excellent match between man and instrument. There he did his great work. Hubble, typical of so many observational astronomers and experimental scientists, disdains speculative thinking, preferring instead a critical skepticism. He remarked:

> Research men attempt to satisfy their curiosity, and are accustomed to use any reasonable means that may assist them toward the receding goal. One of the few universal characteristics is a healthy skepticism toward unverified speculations. These are regarded as topics for conversation until tests can be devised. Only then do they attain the dignity of subjects for investigation.

Through his and his collaborators' work, galaxies, once merely the subject of "unverified speculations," now "attain the dignity of subjects for investigation."

By the late 1920s it was clear that the universe is organized into galaxies of a variety of sizes and shapes, each one consisting of billions to thousands of billions of stars. Between the galaxies was, as far as anyone could determine, black empty space. The galaxies were islands of light in a vast, unending sea of darkness. Why the universe should organize itself this way, rather than as a uniform population of stars, remains a puzzle to this day—a puzzle which, if solved, would shed light on the problem of the origin of the universe, for it is clear to astronomers today that the galaxies are nearly as old as the universe itself.

Hubble's work on galaxies led him to make an important further contribution to an understanding of the expansion of the universe, something on which many people had speculated before him. Like many major discoveries, the discovery of the expansion of the universe did not come all at once. The way was prepared during the years 1912 to 1923 by Vesto Slipher, the American astronomer, who made careful measurements of the shift in the color of light emitted by nearby galaxies. He found that most galaxies had light that shifted toward the red. We know that light emitted by an object moving away from us shifts to lower frequencies corresponding to red, just as a train horn sounds lower as it moves away—an effect called the Doppler shift. So the simplest interpretation of Slipher's "red shift" is that most galaxies are moving away from us—a strange conclusion if one conceived of the galaxies as moving about randomly in the space of the universe.

However, Slipher's observations fitted in nicely with a 1917 cosmological model, based on Einstein's new general relativity theory and invented by the Dutch astronomer Willem de Sitter, which implied that the space of the universe could expand, taking the galaxies with it so that all galaxies move away from each other. For a while Slipher's observations were referred to as the "De Sitter effect." Carl Wirtz, a German astronomer, used this De Sitter effect in 1923. Combining it with rough estimates of the distances to galaxies based on their apparent size, he proposed the velocity–distance law that the velocity of a galaxy (which can be determined from the amount of red shift) is proportional to its distance away from us, thus anticipating Hubble's Law.

Hubble brought his great skills as an observer and measurer of the universe to bear upon the problem of the galactic red shifts. By means of his careful distance measurements, using the Cepheids and M. L. Humason's extension of Slipher's red shift work, he was able to clearly demonstrate that the more distant the galaxy, the proportionately faster its recession velocity, a relation known as Hubble's Law. Since Hubble's time, astronomers have extended their observations of distant objects more than ten times deeper into space, yet Hubble's Law continues to hold today, within the limits of measurement error.

Hubble's Law implies that the universe is uniformly expanding. What does this mean? Since all distant galaxies are observed to be moving away from us, one might think that our galaxy is somehow in the middle of the expansion, and that we occupy the center of a cosmic explosion. This is a misunderstanding. The property of a uniform expansion such as the one Hubble's Law implies is that no matter which galaxy one happens to be in, all the other galaxies are moving away from it; we do not have a privileged location.

Another common misunderstanding about the expansion of the universe is that the galaxies are moving *in* the space of the universe the way a swimming fish moves in a river. The interpretation given to the expansion of the universe by Einstein's general relativity theory is that the galaxies are moving *with* the space, the way a small chip of wood moves with the motion of a river. One may imagine that it is the space of the universe itself which is expanding like a sheet of rubber stretching; the galaxies are just going along for the ride.

According to Einstein, space is not some fixed, immutable entity, a concept in our heads. The geometry and properties of physical space are to be instrumentally determined by what we can measure. But how can one measure the geometry of the space of the entire universe? We must use the galaxies themselves as the "markers" which define that space. If we do that, we find that space itself must be expanding because the galaxies are moving apart. Hubble's Law was thus of great cosmological significance—it was the first empirical law about the structure of the entire universe. All subsequent mathematical models of the universe will have to take it into account.

Hubble devoted many years to classifying galaxies, precisely defining the different types and shapes. Galaxies can be lumped into two main groups, the elliptical and the spiral galaxies. The ellipticals range in shape from nearly spherical arrangements of stars to flat disks. Examples of ellipticals are the two satellite galaxies of the Andromeda galaxy. Ellipticals have a great range of masses and sizes and, interestingly, have almost no gas or dust, just stars. They comprise more than 60 percent of the galaxies, and most ellipticals are small "dwarf" galaxies.

By contrast, spiral galaxies are disks of stars with a central bulge and halo, ranging in visible mass between 10 and a few hundred billion solar masses and harboring lots of gas and dust. They are further subdivided into normal spirals (like our own galaxy and the Andromeda) and barred spirals, which have two bars or jets emerg-

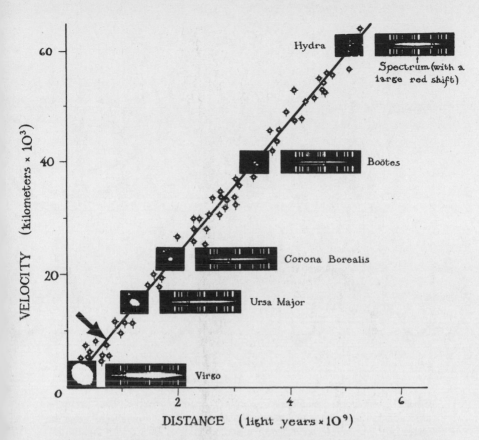

Galactic red shifts (proportional to the velocity of a galaxy) as a function of the distance of the galaxy from us. The linear relation between these quantities is known as Hubble's Law. The apparent size of the galaxy is shown, as are the spectral lines of specific galaxies, indicating the amount of the red shift. The arrow indicates the extent of measurements in Hubble's original study. Since then, the law has been shown to continue to hold (within measurement errors) out to distances nearly ten times farther into space. The data points shown here are illustrative of actual data.

ing from the central region which then terminate in spiral arms (there are no barred spirals in our Local Group, but about one-third of all spirals are barred). Both normal and barred spirals can be further classified in a series, depending on how tightly the arms are wound up.

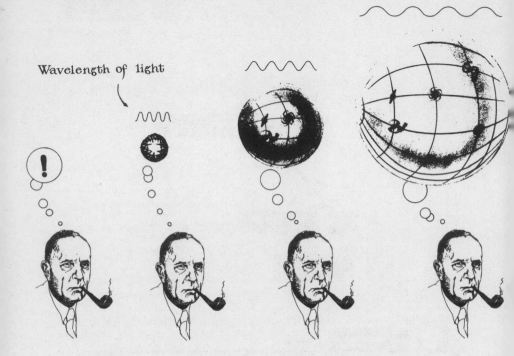

Wavelength of light

Edwin Hubble contemplates the expanding universe. Hubble's Law lent great credibility to the earlier speculations of astronomers and cosmologists that the space of the universe was indeed expanding. Here is illustrated a two-dimensional analogue of the closed FWR universe. As it expands, the galaxies in that space move with the general expansion—the "Hubble flow." The wavelength of a light wave left over from the big bang also increases as space expands. This implies that such radiation in the universe loses energy and, hence, the temperature of the radiation bath of light also decreases. Today, the temperature of the background radiation is a mere 2.7 Kelvin.

Hubble's complete classification system is represented by his famous "tuning-fork diagram." At the fork of the diagram is the lenticular (lens-shaped) galaxy which is like a spiral but without the arms and gas. No lenticular galaxies had been observed at the time Hubble proposed his classification system, but many have subsequently been found.

A few galaxies did not fit into his classification scheme—the irregular galaxies, like the Magellanic Clouds near our Milky Way, which

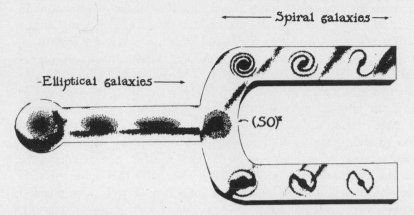

Spiral galaxies →

Elliptical galaxies →

~ (SO)*

← Barred spiral galaxies

*Lenticular galaxy

Hubble's tuning-fork diagram for the classification of galaxies. On the base of the fork are the elliptical galaxies, classified according to their eccentricity. At the joint of the fork is the SO, or lenticular, galaxy. The two prongs of the fork are the spirals and barred spirals. Astronomers once thought this classification represented an evolutionary sequence, but they now believe the different kinds of galaxies are a consequence of different environments and conditions at formation.

Walter Baade called "the wastebasket of Hubble's system." The irregulars are usually observed to be close to an ordinary galaxy. Perhaps they were once ordinary galaxies themselves but got distorted in shape because of gravitational tidal interactions with their neighbor. Baade, who used Hubble's classification system for thirty years, remarked that "although I have searched obstinately for systems that do not fit it, the number . . . is so small that I can count it on the fingers of my hand."

But what is the meaning of the classification system for galaxies? Many of the astronomers, including Hubble, who first reflected on this question, thought the different types of galaxies represented different stages of galactic evolution. They thought galaxies evolved from ellipticals to spirals—left to right on the tuning-fork diagram. In time, some astronomers reversed this and thought the evolution went from spirals to ellipticals—right to left on the diagram.

Today, most astronomers believe the classification system has little to do with galactic evolution and that all galaxies came into existence at about the same time. They know now that all the galaxies contain stars which are some 10 billion years old—good evidence that all the different galaxies are at least this old and thus nearly the age of the universe itself. According to the modern view, different types of galaxies are analogous to the different human races, which are also not an evolutionary series but rather represent different responses to different physical environments. The different kinds of galaxies in Hubble's classification probably reflect differences in the total mass and rotation rates of individual galaxies—physical conditions that were established at their creation. But even today, no one knows the reason for the different shapes of galaxies, why they have the sizes they do or why they form clusters and superclusters. Such puzzles, as well as the related puzzle of the origin and evolution of galaxies, are at the frontier of current research, and we will return to them in a subsequent chapter.

How far away are galaxies? Hubble, through his observations of pulsating bright Cepheids, could establish the distance to nearby galaxies, but the distances to still farther galaxies, where one could no longer see the bright Cepheids, required other methods. One such method is to assume that the very brightest stars in different galaxies all have the same luminosity. Using these as "standard candles" gives astronomers another yardstick. To check this assumption, they can also measure the apparent size of very luminous gas clouds or the brightness of supernovas in distant galaxies to independently help establish the distance. Such methods (they all give about the same results) work out reasonably well to distances of 100 million light-years. Beyond that, astronomers must use the apparent brightness of the galaxies themselves as the "standard candle" because they cannot resolve objects inside the galaxies. But this can be a dangerous assumption. For as astronomers look out deeper into space, they are also seeing the galaxies as they were long ago. If galaxies are evolving, then perhaps their light output was greater in the past and not the same as that of the older galaxies we see nearby. The method could thus lead to erroneous estimates of the distances.

An example of how much our measurement of astronomical distances depends upon our assumptions regarding the luminosity of distant objects is Walter Baade's discovery during World War II of two populations of stars. Here is the story of how Baade's work implied that the universe was twice as big as people had previously thought.

After Hubble's work on the Cepheid variables in nearby galaxies, the distances to our neighboring galaxies were thought to be established. But on the basis of these distances, astronomers could also determine the absolute size of other galaxies, and it seemed that our galaxy was about twice as large as other spirals—a complete anomaly in its class. This anomaly disappeared after Baade showed in 1952 that the Cepheids used by Shapley to determine the distances to the globular clusters and hence the size of our Milky Way galaxy were the older population II stars, while those Cepheids observed by Hubble in the Andromeda galaxy were the bright young population I stars in the spiral arms. The old and young Cepheids from the two star populations were physically different and not surprisingly had different period–luminosity relations. Hubble's distance estimates to the galaxies, which did not take the distinction of the two star populations into account, turned out to be too small. After Baade's important work, the distance scale of the universe was increased by a factor of 2, and as a consequence our galaxy is no longer anomalously large but a typical-sized spiral. There is order in heaven.

It is difficult to estimate the distances to the farthest galaxies reliably. Yet it is extremely important to know those distances, in order to map out the large-scale structure of the universe. Until we do this, our current map of the universe, or even of the interior of our own galaxy, can be likened to those maps of the New World or Asia drawn by cartographers from reports made by the first European navigators—a mixture of observation and guess. Using new instruments, like the space telescope, astronomers will begin to construct more reliable maps of the cosmos.

When it comes to learning about the details inside a galaxy, there is no better place to begin than at home, with our own Milky Way galaxy. What, then, is our own galaxy—and by implication other spiral galaxies like it—made of? The two main visible components of our galaxy are stars and the interstellar medium, which, although it has only 5 percent of the mass of the stars, is very important to a determination of the dynamics of the galaxy. The interstellar medium consists of gas, dust, cosmic rays (high-energy quantum particles) and magnetic fields. All these components—plus perhaps other unknown ones such as dark matter in the halo—make up our galaxy and interact in a complicated, not yet fully understood way. The galaxy is a rich, chemically changing, dynamically evolving system, every bit as complicated as the weather on earth. Let us examine these components of the galaxy in turn and see what astronomers have learned.

The stars, the most conspicuous component, can be classified according to their spectral type, surface temperature, luminosity, mass, chemical composition and age—a variety of characteristics which are interrelated and reasonably well understood. In our galaxy the stars are thickly concentrated near the nucleus and thin out in the disk. Older stars are found throughout the disk, while the younger ones like our sun tend to be near the galactic mid-plane. All the disk stars are moving in complicated trajectories, as well as oscillating up and down through the galactic mid-plane. They revolve in one direction about the center. Astronomers estimated the rotational velocity of our sun about the center of the galaxy to be about 200–300 kilometers per second. A few high-velocity stars actually fly out well beyond the disk and spend most of their time out in the halo. These solitary halo stars are known to be quite old and may be escapees from the globular clusters that populate the halo, or perhaps they are the original stars from which the globular clusters themselves were made.

The first solid evidence for the existence of an interstellar medium came in the 1920s. Astronomers could observe the medium because of its characteristic absorption of starlight and concluded that clouds of ionized calcium and neutral sodium atoms were floating around between the stars. But astronomers reasoned that since stars were made mostly of hydrogen, if an interstellar gas existed, it ought to contain large amounts of neutral hydrogen gas as well. The trouble with verifying this line of reasoning was that unlike calcium and sodium atoms, which absorb light if cold, only hot hydrogen atoms would emit visible light. Since any hydrogen atoms in interstellar gas clouds would be very cold indeed, astronomers could never even see them in spite of the fact that they might be very plentiful.

The way to observe the cold hydrogen clouds was pointed out by the Dutch astronomer H. C. van de Hulst during World War II. Jan Oort, the director of the observatory at Leiden, had read the articles of the pioneer American radio astronomer Grote Reber and reported on them to his group in 1944. Oort emphasized the importance of detecting the neutral hydrogen, and Van de Hulst picked up on this. He knew that neutral hydrogen consists of a single electron in orbit about a single proton. The proton and the electron each have a unit of spin—one can imagine them to be little spinning tops—and hence, cold neutral hydrogen can have the electron and proton both spinning in the same direction or in opposite directions. There is a small energy difference between these two spin configurations, and when the hydrogen atom makes a quantum jump from the higher-energy state to the lower, it emits radiation. The wavelength of this radiation can be calculated to be 21 centimeters—just the size that can be detected by a properly designed radio telescope.

After the war, when astronomers built such radio telescopes, the huge clouds of neutral hydrogen in the disk and especially the arms of the galaxy were "seen" for the first time. The 21-centimeter radiation provided a new way of mapping the galaxy, and one that supported the earlier conclusion that it was a spiral. Over the last fifteen years astronomers using the new X-ray satellites, as well as infrared and radio telescopes, were surprised to discover that the interstellar medium is distributed not in clouds of roughly spherical shape scattered among the stars, as they had previously thought, but rather in huge expanding shells of gas. Such shells of gas might be produced by a gigantic stellar wind moving away from a hot star or a supernova. Radio astronomers also found evidence that some gas might lie well outside the plane of the galaxy, forming a galactic corona.

Besides gas, our galaxy also contains dust—small particles of matter perhaps one one hundred-thousandth of a centimeter in size. No agreement exists about the chemical composition of this granular dust; it could be silicates, graphite or dirty ice crystals. The dust absorbs starlight (the blue more than the red), so the stars behind the dust appear dimmer than they would otherwise. It is important for astronomers to know about the absorption properties of the dust if they are going to calibrate distances to different parts of the galaxy.

Besides absorbing starlight, the grains of dust also scatter it. When such light passing through dust is observed, it is often slightly polarized—its wave oscillations occur in preferential rather than random orientations. This means that the elongated grains of dust which absorb or scatter the light must also be oriented. The most likely explanation for such a uniform orientation is the existence of a galactic magnetic field penetrating the clouds of dust. Like countless numbers of little compass needles, the grains align with the galactic magnetic field. From polarized starlight astronomers first learned that the interstellar medium has a magnetic field.

Independent evidence for galactic magnetic fields comes from the observation of cosmic rays—high-energy elementary particles raining down upon the earth, uniformly from every direction in space. These cosmic particles may be produced either freshly from supernova explosions or, as seems more likely, from accelerated interstellar material (the relative abundances of the elements in cosmic rays agree with those of the interstellar medium). But because the cosmic particles have such high energy, they should fly right out of the galaxy after they are produced and we should observe very few of them. Instead, we detect lots of cosmic rays. Something has to hold these high-energy particles inside the disk of the galaxy, and the only viable explanation is the existence of a galactic magnetic field. Magnetic fields can "bottle up" electrically charged particles, and that is just what the galactic magnetic field does; it seems to be a permanent fixture of our galaxy.

Besides mapping out the interstellar hydrogen clouds, radio astronomers, joined by infrared astronomers, discovered the existence of interstellar molecules, which emit radiation near the famous 21-centimeter hydrogen wavelength. These discoveries marked the coming of age in the late 1960s and 1970s of the new methods of radio spectroscopy, methods which enabled astronomers to study radio-spectral lines indicating the presence of molecules just as they previously had studied visible-spectral lines indicating the presence of atoms. In this way, carbon monoxide and formaldehyde molecules, among others, were discovered. But astronomers were completely surprised by the discovery of large organic molecules such as ethyl alcohol and cyanoacetylene in a small number of dense clouds. What were more than fifty different such big molecules doing in deep space? These large molecules will form from simpler ones and remain preserved only under rather special temperature conditions. Such conditions are present in the contracting warm clouds of gas that make up the giant molecular cloud complexes discovered in the 1970s, which will almost certainly go on to form new stars. So the existence of these large molecules gave astronomers some more clues toward understanding the complex star-formation process.

Stars are born in the dense gas clouds found in the spiral arms of the galaxy. But what is the origin of the arms? If the spiral arms always consisted of the same stars revolving together in the disk, the stars near the center would revolve more rapidly than the outer stars. If this picture is right, then the spiral arms would get wound up tight around the galaxy in a few galactic rotations taking several hundred million years. But this is not what is seen—the arms preserve their shape as the galaxy rotates.

To resolve this "winding dilemma," the Swedish astronomer Bertil Lindblad suggested a new density-wave theory in 1941 which was further developed in the 1960s by the American mathematician Chia Chiao Lin and his associates. According to this theory, a spiral-shaped wave in the density of stars in the disk is gravitationally self-sustaining. New stars and dust are continuously being swept into the spiral density wave, just as particles of water in an ocean wave of constant shape are continuously changing. Although this theory resolves the winding dilemma, it leaves other problems—it does not explain why the arms do not get washed out after a few galactic rotations.

In an alternative theory developed by Alar Toomre and his collaborators, the spiral density wave is generated by complicated gravitational tidal interactions with nearby galaxies. That such tidal forces between galaxies can form the spiral structure has been demonstrated by computed simulations of two galaxies passing near or through each other. This picture suggests that the large number of spirals we see today is due to a high rate of dynamical interactions among galaxies throughout their natural history.

Yet another theoretical model of the spiral arms, advanced by Phillip Seiden and his coworkers, incorporates the new ideas of star formation in molecular dust clouds. When massive stars are formed in dense clouds, they generate shock waves in the cloud which, in turn, cause the dust to concentrate further, engendering more star formation. As bursts in the star-formation wave move outward from the center of the galaxy, the rotation of the galaxy turns them into the spiral arms. When put on a computer, this model of "stochastic star formation" generates impressive pictures that look just like real spiral galaxies.

But none of these theoretical models for the origin of the spiral arms is widely accepted. It has been more than a century since Lord Rosse first saw the many spiral nebulae with his telescopes and yet, as Alar Toomre recently summarized, "It seems clear now that the spiral structure of galaxies is a complex riddle, without any unique and tidy answer." The interactions of the galactic magnetic field, cosmic rays, the dust, gas and stars, all bound by gravity, are very complicated. But the dynamics of those interactions, being unraveled by astronomers today, is ultimately responsible for the structure of our galaxy and the processes that give birth to stars.

Yet even as astronomers struggle to understand the processes they observe in our own galaxy, they are looking ever farther out into the depths of space and discovering new objects—radio galaxies, quasars, possibly even black holes. The astronomers have progressed far beyond the eighteenth-century speculations about the nebulae. But major exciting questions remain—questions about the origin and evolution of galaxies, their internal dynamics and generative processes, the nature of the mysterious nucleus of the galaxy. The fact that such challenging questions remain unanswered guarantees that galactic astronomy will remain an open and exciting field for decades to come.

Astronomers during the first half of this century discovered the galaxies—the major organizational feature of our universe. The instruments that dominated this period were the large optical telescopes used in conjunction with modern photographic technology. For a long time that seemed to be the only way of doing astronomy. But during and after the Second World War scientists developed radar technology and its offspring the radio telescope (already deployed before the war), which opened another window on the universe.

The war was a watershed in the sciences, resulting not only in new astronomical techniques but also in a social reorganization of scientific laboratories. Prior to the war, observing time on the large telescopes was assigned by the observatory director, usually a benevolent dictator with a style all his own. After the war, the decision making shifted to a more complex and democratic method, an openly competitive system. This destroyed the quiet life atop the mountain with the telescope. As Jesse L. Greenstein, a distinguished astronomer and staff member of the Mount Wilson and Mount Palomar Observatories since 1948, remarked, "The world changed. The characters in the play changed. Instead of the gentlemen, we have as the ideal the brilliant, aggressive young genius interested in everything, careless of whose feet he steps on and very anxious to make the discovery of the week or the year or whatever. It's a loss, and, it's a gain." In the late 1940s, the publications of astronomers increased, the administrative burden of the large observatories grew and human knowledge of the cosmos expanded.

Human consciousness was extended into the depths of space, the cores of stars and the beginning of time. Scientists asked new questions and repeated old questions, questions which for the first time acquired empirical content. In the next chapter we continue our exploration of the realm of the galaxies, especially the revelations from the radio telescopes, instruments that are opening a new universe for astronomical science.

5
Radio Galaxies and Quasars

If the depth of the universe was to be probed, larger and more sensitive instruments were required. The realization of large radio telescopes is a story of scientific vision and struggle, competition and controversy, persistence and diplomacy.

—J. Stanley Hey

George Ellery Hale devoted the last years of his life to building the largest reflecting telescope. The heart of this telescope was an immense light-gathering precision mirror, 200 inches in diameter, which could resolve faint objects in the sky with exposures of reasonable duration. A second "Schmidt telescope," a new device with a special lens before its mirror that could take in a tremendous area with minimal distortion near the edges, was to work in tandem with it—a powerful pair exploring the heavens. Hale saw the great 200-inch mirror arrive in California shortly before he died in 1938.

After World War II, it went into operation on Mount Palomar, California, to become the major optical instrument for the next several decades. Appropriately, it was called the Hale Telescope. Hubble proposed using it to continue his work of counting galaxies and expected to get most of the best observing time during the dark of the moon. A meeting of the leading astronomers was held in a private house one afternoon in 1948 in Pasadena, California, to decide the issue of allocating observing time. As Hubble listened, his colleagues Richard Tollman, Walter Baade, Ira Bowen and others explained to him that the counting of galaxies was not going to yield useful information about the expansion of the universe. Hubble, whose prestige and presence had done so much to help raise the funds to build the telescope, would not have his proposal realized. While profoundly disappointed, Hubble accepted the decision like a gentleman.

Even while optical astronomers were making great strides in exploring the universe with their new telescope, the whole new field of radio astronomy opened up. Back in 1931, Karl Jansky of the Bell Telephone Laboratories first detected radio noise—"This hiss from the depths of the universe," as a radio announcer called it. Eight years later Grote Reber, a radio engineer, became the first radio astronomer. Using his savings, he built a radio dish reflector 31 feet in diameter in his backyard in Wheaton, Illinois—a project which in its prospects resembled Herschel's building his first reflecting telescope in Bath, England. Reber's radio telescope had sufficient directional signal-detection capability for him to determine that the most powerful signals came from the center of the Milky Way and from two sources in the constellations Cygnus and Cassiopeia, respectively. Reber went on to make the first rough radio map of our galaxy.

But astronomers paid little heed to Reber's observations. The interests of astronomers and radio engineers seemed so disparate that the great prospect of radio astronomy was not grasped until later, especially in the United States. Radio engineers were interested mainly in those radio waves which could be bounced off the atmosphere for long-distance signal transmission, not waves like those Reber studied which penetrated the atmosphere. Astronomers for their part were simply not interested in radio waves. As Reber remarked in 1948 when he attempted to get support for a larger radio telescope, ". . . for the most part the attitude was that I was harmless and if no interest was shown in what I was saying, I would go away quietly."

Some astronomers thought that radio astronomy, because the wavelengths of radiation involved were so long, could not compete in resolving power with optical astronomy and its much shorter wavelengths of visible light. What was not realized was that if several radio telescopes were placed far apart the effective aperture and hence resolving power could be greatly increased. But the main reason these early leads were not followed was that no one could envision emitting mechanisms for the radio signals other than ionized gas, and that was rather localized in spots in the galaxy and not especially interesting. But it has turned out the universe is richly populated with astronomical radio sources which involve other mechanisms that no one anticipated back in the 1940s.

The big breakthroughs in radio astronomy occurred after World War II when scientists gained the experience of using new electronic technology like radar. A new breed of scientist, equally at home in radio engineering, radar and astronomy, became actively involved in constructing the big radio dishes, "antenna farms" and auxiliary electronic equipment required to do the new astronomy. Although scientists like Jansky and Reber in the United States had done the prewar pioneer work, it was the British, Dutch and Australians who followed their lead and made most of the major early discoveries. The often foul weather conditions in England did not inhibit radio waves as they did ordinary starlight. At Cambridge University, then the center of science in England, J. A. Ratcliff, a major figure in radio and ionospheric research, encouraged inventive young radio engineers like Martin Ryle to enter radio astronomy. Ryle first linked together several small radio dishes to produce the resolving power of a single large one. Another group, led by Bernard Lovell at Manchester, built the Jodrell Bank radio telescope; at 250 feet in diameter it was the first of the big dishes. Different radio facilities in Australia, the Mill's Crosses (each an antenna arranged like an immense cross on the ground), listened in on the southern skies. Radio astronomy progressed rapidly.

Although radio astronomers had some successes in the 1940s, the big payoff for their labors came in the years 1950–1951. The major discoveries radio astronomers made in those two years marked a turning point and finally established radio astronomy as a major new scientific discipline providing profound insights into the physical nature of the universe.

During these two years, astronomers determined that the strong radio source Cassiopeia A was actually a supernova remnant—a ring of radio-emitting gas—that marked the place where a star had exploded around A.D. 1700. Cygnus A, another radio source, proved to be a peculiar double galaxy a million times as powerful as an ordinary galaxy in its radio output. These years also saw the first radio map of an external galaxy, the Andromeda galaxy. As if these discoveries were not enough, three groups of scientists at Harvard, in the Netherlands and in Australia detected the 21-centimeter neutral-hydrogen radiations predicted by H. C. van de Hulst back in 1945, thus finding a whole new method of mapping our own galaxy. Since the Doppler shifts in the radio wavelengths—essentially the red and blue shifts—of the moving clouds of neutral hydrogen could be measured even more accurately than the Doppler shifts of visible light, radio maps ultimately provided the greatest detail of the spiral structure of our galaxy.

Even while these observational discoveries were being made, in 1950 the Swedish theoretical physicists H. Alfvén and N. Herlofson first suggested that the radio emission from discrete radio sources originated as synchrotron radiation—a form of high-energy electromagnetic waves caused by electrons gyrating around lines of magnetic fields—thus providing an explanation for many radio sources. Electrons, and indeed all electrically charged particles, will radiate electromagnetic waves if they run around in little helices, and the rapidly moving, energetic electrons in a magnetic field do just that. Today, in fact, synchrotron radiation is observed in earthbound particle-accelerator laboratories called synchrotrons (hence the name), and the resulting radiation is used for a variety of therapeutic medical purposes. So the mechanism for generating the radio waves seen by radio astronomers is now well understood.

Responding to these dramatic developments in 1950–1951, J. Stanley Hey, an English radio astronomer who first discovered radio waves from solar flares, remarked:

It was by now clear that radio astronomy could make a great contribution to astrophysics. The fusion of optical and radio astronomy had already commenced. . . . Much sustained observational work lay ahead. . . . If the depth of the universe was to be probed, larger and more sensitive instruments were required. The realization of large radio telescopes is a story of scientific vision and struggle, competition and controversy, persistence and diplomacy.

A new generation of astronomers, continuing the work of the past, came to the fore with powerful new instruments at their command. The training of astronomers also changed, reflecting the new needs. In the decade following World War II, only a small minority of astronomers were trained as physicists; the rest were trained as traditional astronomers. Today, anyone seriously interested in pursuing astronomy has to have a background in physics—nuclear, elementary particle, plasma and atomic physics in particular—and 35 percent of doctoral-level astronomers have their Ph.D.s in physics.

The basis for the success of radio astronomy also became clear to all scientists. Because they are less energetic than ordinary light waves, radio waves are easily produced by the natural processes that occur in astronomical objects, and hence the universe abounds in radio sources, just as it does in sources of visible light. A second reason for the success of radio astronomy is the incredible detection efficiency of the large radio telescopes. Because the radio-signal detector at the focal point of the telescope can be artificially cooled to very low temperatures, the background noise that interferes with the signal can be made quite small; hence the radio range of frequencies is the most sensitive part of the entire electromagnetic spectrum. Such sensitivity provoked the admiration of scientists like Robert Otto Frisch, who once commented:

When the Mullard Radio Astronomy Observatory was inaugurated here [Cambridge, England] in 1958, each guest at the luncheon found a little white card by his place, with these words on the back: "In turning over this card you have used more energy than all the radio telescopes have ever received from outer space."

Not only could no one have anticipated such sensitivity, but also no one thought that radio astronomers could achieve a resolution of a thousandth of a second of arc—equivalent to resolving a dime at a distance of 4,000 kilometers. The resolving power of telescopes is related to the size of the detector. Radio telescopes, unlike optical telescopes, can have their detectors spread out over an immense area on the face of the earth and then have the individual signals combined. With such incredible resolving power it comes as no surprise that it is radio astronomers who can resolve the most distant objects in the universe.

In the decades following the great successes of the early 1950s, radio astronomy underwent a twofold expansion: on the one hand scientists built large single-dish reflectors and on the other hand they deployed long-baseline interferometers (LBI) in which multiple, but spacially separated, radio detectors mixed together signals from the same source. The United States got into the new field, as did the Soviet Union, and today most radio telescopes are in the United

States. The most recent large radio telescope is the Very Large Array (VLA) near Socorro, New Mexico, consisting of 27 linked radio telescopes, each with a dish 25 meters in diameter. The baseline of the VLA is about 17 miles. Yet another technique, consisting of linking up separate radio telescopes on different continents, is called very-long-baseline interferometry (VLBI) and provides the highest resolution of any method. These new instruments explored the radio galaxies and enabled radio astronomers to meet their colleagues in optical astronomy on equal terms for the first time.

The work of radio and optical astronomers was often complementary, as it was in mapping of the visible surfaces of the moon and planets, examination of the solar chromosphere and observation of flare stars. But many major discoveries could have come only from radio astronomy—among them the galactic distribution of neutral hydrogen; the mapping of the magnetic fields of galaxies; the discovery of radio galaxies, quasars, pulsars, molecular emission lines, the interplanetary plasma and the Jovian radiation belts. Such discoveries created a new view of the universe which indicated the presence of complex dynamical processes. A previously invisible universe was rendered visible.

Radio astronomers found hundreds of radio sources in the heavens, and in many cases these sources could be identified with optical counterparts. Some of these were local sources in our own galaxy, such as the gas cloud associated with the Crab nebula—a supernova remnant. But the most powerful optically identified radio sources, with intensities thousands, if not millions, of times the radio output of our own galaxy, were a new class of galaxies—the radio galaxies. What one sees are chaotic objects, giant elliptical galaxies with brilliant central regions or long jets of ejected matter protruding from their cores. They are clearly very active galaxies. Yet most radio galaxies, blazing away radio radiation, cannot even be seen optically—they are too far away and dim in their output of ordinary light. Yet these "invisible" galaxies could yield powerful clues to a solution of the puzzle of the evolution of the visible galaxies.

With the construction of high-resolution radio telescopes it became clear that extragalactic radio sources sorted themselves out into two categories—"compact" and "noncompact" sources. Most radio sources are double; the radio radiation from a single regional source comes from two distinctly separated components. In the compact sources the two components are only about a hundred light-years

A radio-intensity plot of an active radio galaxy (at the center) ejecting "fountains of radio stuff" into space. Two jets of matter emerge from a central object, eventually losing energy and ballooning out to form two large lobes.

apart—a relatively short distance if we recall that our galaxy has a diameter a thousand times as great. But some double sources are "noncompact"; they occupy a huge region of space and their two lobelike components are separated by thousands, if not millions, of light-years. But what causes the lobes?

The radio-emitting lobe-shaped regions are due to long "cosmic jets" of streaming plasma (ionized gas) from a central core. The jets, what Philip Morrison of M.I.T. calls "fountains of radio stuff," would sometimes be bent and twisted in space, ending in hot spots —the lobes of intense radio emission. Sometimes only one jet could be seen. It was found that the large-scale jets were generally associated with low-intensity double sources; the feeble radio galaxies have two opposing jets, while the powerful ones have none or only one.

All the evidence suggests that the jets are streams of gas propelled from the center of the radio galaxy like water from the nozzle of a hose. The supersonic gas jet passes through the interstellar medium of the radio galaxy and encounters the intergalactic medium of thin gas, becoming slowed down as it pushes against it, even producing a hot shock front. The lobelike regions are those hot spots where the jet slows down and energy accumulates. After the matter from the jet has been slowed down it gradually flows back to the galaxy, thus inflating the large lobes seen by the radio astronomers. The bent and twisted jet stream is just the streaming gas of electrons and other charged particles as it winds its way through the intergalactic medium.

The central puzzle of radio galaxies is Where do those electrons come from in the first place—what produces the jets in the nucleus of the galaxy? None of the usual ideas of physics seem able to account for the creation of all those energetic electrons. Astrophysicists have speculated on possible mechanisms for the sources of the jets, and I will describe a few of their ideas. But before plunging into these speculations, I will describe what to many people was an even more surprising discovery than that of radio galaxies: the discovery of quasars, a new class of astronomical objects. Like radio galaxies, quasars were first discovered by the radio astronomers, but much of the detailed study of these new objects was made by optical astronomers—evidence of a fruitful collaboration between the two branches of astronomy. But what are quasars?

Most of the hundred or so radio galaxies spotted up until the early 1960s appeared as fuzzy objects. The optically identified ones were clearly giant elliptical galaxies. But a few of the visible radio sources were not fuzzy but starlike in appearance and also emitted an unusual amount of ultraviolet radiation. They seemed different, and indeed they were.

The breakthrough came in 1963, when Maarten Schmidt of the Hale Observatories first identified emission lines in the light spectrum of one of these starlike objects—3C 273. The spectral lines were shifted into the red to the incredible degree of 16 percent. Such a red shift, if due to the expansion of the universe, meant that the object was about 2 billion light-years away—very far away indeed. These starlike objects, given the name quasars—"quasistellar radio sources" —are now generally regarded as among the most remote objects in the universe and are emitting immense amounts of energy. After the quasars were identified, Allan Sandage went on to discover many radio-quiet quasars with the 200-inch Hale telescope on Mount Palomar. He noticed that quasars emit much more ultraviolet light than ordinary stars and hence could be optically picked out of the multitude of pointlike sources of light.

To get a feeling for the energy output of a quasar, imagine that a galaxy is the size of a room. A quasar would not be bigger than a barely visible speck of dust. Yet a single quasar produces 100 times the energy radiated by all the billions of stars in our galaxy. Now one can appreciate how incredible the energy output of a tiny quasar is! And it radiates this energy for about 10 million years—a total amount equivalent to converting 100 million suns into pure energy.

What compounds the puzzle of the quasar is that its active region is so small—not much bigger than about one light-week, or about 10 times the size of the solar system. How do astronomers know the size of something so distant? Since nothing travels faster than light, in order for there to be coherent fluctuation in the intensity of light over some region of space, the size of that region cannot be greater than the distance traveled by a light signal during the duration of that fluctuation. Quasars are observed to alter their intensity over periods as short as a week, and hence the source must be smaller than one light-week across.

Like radio galaxies, quasars have both compact and noncompact radio-emitting regions. The less-brilliant quasars resemble the nuclei of the brightest radio galaxies, and the least-active radio galaxies resemble ordinary galaxies. This continuous sequence in the activities of galaxies suggests to many people an evolutionary sequence in which quasars, of which there were many more in the past, evolve into radio galaxies which, as they quiet down, turn into ordinary galaxies—an evolution from high activity to low or sporadic activity in the core of a galaxy. But this possible evolutionary sequence, while intriguing, is still currently being debated by astronomers. It could very easily be completely wrong, or only part of the story of galactic evolution. Yet it seems attractive.

What mechanisms within the laws of physics could possibly power the quasars or radio galaxies and produce the observed jets? To answer this question we leave the realm of observation and turn to the speculative imagination of the astrophysical theorists. They think that the core of a galaxy harbors a "monstrous machine" which may involve one or more black holes or similar compact objects. Edwin Salpeter from Cornell University and Yakob B. Zel'dovich, a Soviet physicist, are among those who espouse this view. They theorize that in a very early stage of the formation of a galaxy, when it consists mostly of ordinary stars and gas, one star or more in the densely packed core of the galaxy explodes, converting itself into a black hole. Then, like a cannibal eating its way through a population, the black hole consumes millions of its densely packed neighbors, swelling enormously to 100 million solar masses—a process that takes a few million years. Although the black hole cannot itself radiate, the gas that is pulled into the hole does. Physicists estimate that one-tenth of the mass sucked into a black hole is converted into radiant energy and this radiation is what we see as a quasar. If the black hole consumes about a billion stars the size of the sun, then the radiant energy is equivalent to 100 million stars—the right amount to explain the energy output of the tiny quasar.

Black holes are also popular with theorists attempting to explain the behavior of radio galaxies. According to their models, the electrically charged gas as it swirls about the hole forms an electric dynamo which projects two jets of high-energy particles in opposite directions from the hole. Black holes can rotate, and the jets emerging along their axis could be fueled by the rotational energy. These are then the "fountains of radio stuff" seen by the radio astronomers.

Another speculation is that the radio-galaxy core contains two closely orbiting black holes, or a supermassive star in orbit about a black hole, or even just two supermassive stars. The jets emerge from the interactions of the two massive objects as they orbit in the nucleus of the galaxy.

Jeremiah Ostriker of Princeton University and Scott Tremaine, now at M.I.T., have formulated another intriguing model. They envision that the black hole of a large "cannibal" galaxy consumes a smaller "missionary" galaxy, tearing its outer parts to pieces and, in the process, capturing in orbit the missionary's black hole. Part of this idea's attractiveness lies in the fact that most radio galaxies existed in the distant past, when galaxies were more densely packed and "cannibalism" would have been more likely. Also, several features of the observed jets can be explained by this hypothesis. But there are lots of theoretical models for explaining the radio galaxies, and it is hard to determine which, if any, are correct.

Suppose these theorists are right and there are black holes in the cores of quasars and radio galaxies. What happens after the black hole has done its work? After a black hole has swept the core clean of most stars and gas, it quiets down, the radio jets atrophy and the galaxy becomes a normal galaxy like most of those in our cosmic neighborhood. Probably the black holes at the centers of spiral galaxies are much smaller than those in ellipticals, because the angular rotation of the stars in spirals keeps them out of the dangerous core region.

If these ideas are correct, then there may indeed be a black hole of a few million solar masses at the center of our galaxy, as conjectured more than a decade ago by the English astrophysicists Donald Lynden-Bell and Martin Rees. The best evidence for this idea comes from recent radio observations of the galactic center which reveal a nonthermal compact radio source (presumably the black hole) surrounded by spiraling arms of gas, perhaps originating from torn-apart stars. The radio signal astronomers see is just the kind that is expected to be detected for gas falling into a black hole. Further evidence is provided by X rays and gamma rays which can also be observed coming from the galactic center. They result from electron–positron annihilation, a process that could take place just outside a black hole.

Until more data become available, this black-hole model of the galactic core will remain in the twilight zone of speculative physics. Fortunately, much more data will be gathered in the next decade. Some will come from the Very Large Array in Socorro, New Mexico, which so far has mapped out only a fraction of the radio sources it can see and only a few of these with high resolution. The use of this important instrument is only beginning, and already it has revealed new features of the galactic core. The VLBI program, which simultaneously utilizes radio telescopes located on different continents, will also be enhanced. The space telescope, with 20 times the resolving power of comparable ground-based telescopes, will tell us a lot about quasars, and we may discover many more optically visible jets. One interesting question that might get answered by the Space Telescope is whether or not quasars are located in clusters and super-clusters as are galaxies.

Recently astronomers discovered "gravitational lenses," entire galaxies that are situated on the line of sight between the earth and a distant quasar. The galaxy itself may be invisible, but its immense mass curves the surrounding space, causing light from the distant quasar to bend around it, producing multiple images of the quasar here on earth. Astronomers are searching for more of these gravitational lenses (one has been found by a direct search) because they may provide new methods for establishing distance scales to the faraway quasars and deriving new information about the large-scale structure of the universe. Gravitational lenses could become "optical benches" on a cosmic scale.

Today astronomers debate the merits of prospective and existing technology such as the multiple-mirror telescopes, which instead of having one large mirror have many smaller ones that electronically combine their separate images into one image. A different design, still in the proposal stage, is a 15-meter segmented-mirror telescope (as contrasted with the old 5-meter Hale telescope) which will look like a polished honeycomb; each of its sixty hexagonal pieces will be individually controlled so that they will form a common focus and present one image. It would be a national telescope. The University of California plans to build a privately financed 10-meter segmented-mirror telescope with thirty-seven hexagonal pieces each about 2 meters across, which would penetrate twice as deep into space as any existing telescope. Astronomers also discuss the use of deformable mirrors that utilize new imaging systems. Many of these new design features are provoked by a friendly competition among telescope builders and are intended to make the telescopes efficient, inexpensive and able to peer farther out into the cosmos. No doubt new components of the cosmos will be revealed by these supergiant telescopes planned for the 1990s, instruments that today seem like devices out of science fiction.

Many of these new instruments will be devoted to the study of very distant galaxies and quasars, at what Allan Sandage calls "the edge of the world." Such a study is interesting because it may answer some of the outstanding problems confronting the modern astronomer: what processes control the evolution of galaxies, and what is the large-scale structure of the universe? But in spite of the technological advances in detection systems made so far, astronomers are frustrated in finding answers to these questions because the data they obtain are not easily interpreted. For example, some distant galaxies have colors inconsistent with their red shifts—an indication that we do not really understand the composition of these galaxies. A further problem arises if we recognize that determining the distance to faraway galaxies requires "standard candles"—a class of galaxies all of which have approximately the same luminosity over time. But if galaxies are evolving, their luminosity may have been far greater in the past and no "standard candles" may exist at all. Thus the problems of determining the evolution of galaxies and the large-scale structure of the universe are intertwined.

Yet another interesting question on which the new instruments might shed light is How old are the galaxies? This is not an easy question to answer because if galaxies like ours existed more than a few billion years ago, they would be too dim and small to see with our present optical telescopes. But quasars did exist, and it may be that they were just the luminous cores of new galaxies, from which ordinary galaxies evolved. When the Space Telescope goes into operation, we ought to be able to check this idea. But for now, assuming ordinary galaxies evolved from quasars, the question of the age of galaxies is then the same as the question How old are quasars?

The oldest (and hence also the most distant) known object in the universe is a quasar discovered by means of one of the new giant telescopes—the Parkes Radio Observatory in New South Wales, Australia—and catalogued "Parkes 2000-330." It is moving away from us at 90 percent of the speed of light, and the light we see from it originated when the universe was only one-fifth its present age. Most quasars are observed to lie in an epoch of one-third to one-fifth the present age of the universe—the era of quasars. Astronomers have looked at earlier times but found few if any quasars. Early ones may be obscured by dust, but it appears as if the quasars turned on and then shut off at about the same time in the history of the universe. If galaxies originated as quasars, then according to some estimates these quasars first turned on 10 to 11 billion years ago—an acceptable time scale compared with the age of the oldest stars and nearly the age of the universe itself.

What did galaxies look like before the era of quasars? Astronomers are searching with new sensitive instruments for "primeval galaxies," some of which may surround quasars but are too dim to be seen. They are believed to have existed from the first 100 million years to the first few billions years after the big bang. Discovery of such a galaxy would be like finding a fossil never before seen. Are they diffuse structures with low surface brightness, or are they blue, indicating a high rate of star formation? What did the universe look like then? No one knows for certain—there are few or no data on this primordial period, and speculation is open to the almost unconstrained imagination.

We have come a long way from Hubble's view of galaxies as isolated island universes. Galactic theory and observation are opening a new view of the evolution of galaxies. Just as many stars within a galaxy form groups and create a rich environment, so do galaxies tend to congregate in clusters and provide an environment within the universe, interacting in complex ways. Elliptical galaxies are more often found in the high-density parts of galactic clusters, while spirals are found to be relatively isolated or in the low-density parts of clusters. There is mounting evidence that galaxies near the center of clusters once were spirals that were somehow stripped of their gas, so that star formation in such galaxies was stopped. The galactic environment is thus important to determination of the structure, shape and evolution of a galaxy.

Stars evolve, galaxies evolve and clusters of galaxies also evolve. The mechanisms of this evolution are only beginning to be understood. Formidable obstacles stand in the way of our enlightenment. We may not even know what the primary material components of the universe are. Invisible forms of matter—black holes on the macroscope scale, new quantum particles on the microscopic scale—may be very important for understanding these processes.

But now let us expand our view of the cosmos beyond galaxies, radio galaxies and quasars. Let us imagine that galaxies are just particles distributed like sand throughout the space of the cosmos. Are they distributed at random or is there a structure—a further clue to cosmic order? Such questions bring us to the problem of determining the large-scale structure of the universe, a topic to which we now turn.

6

Why Is the Universe Lumpy?

There are . . . strong divisions of opinion on how galaxies formed. That is not a sign of trouble, for it is only recently that the subject has advanced to the point that we can make out positions that seem defensible.
— P. J. E. Peebles, 1984

Have you ever watched rain begin to fall on an open pavement? At first there are a few drops wetting the ground; then more fall until the entire pavement is wet. But before that happens, the raindrops can be seen to form a random two-dimensional pattern on the pavement—the spot where one drop falls is independent of whether or not previous drops fell there.

As another example of a random pattern, consider the nesting habits of certain seabirds on an island. Each bird has its place, and if we take a bird's-eye view of the whole nesting group, the birds appear to be located at random in a two-dimensional pattern. But it is a different kind of randomness than what we encountered with the falling raindrops. The territorial birds don't seem to like each other.

OPPOSITE:
Different kinds of randomness characterize raindrops falling on the pavement and nesting seabirds. The distributions are different because raindrops hit a spot on the ground independently of whether a raindrop has previously hit that spot, while the territorial sea birds keep at least a pecking distance between nests. The correlations of the random distribution of galaxies in the universe may provide clues to their evolution and origin.

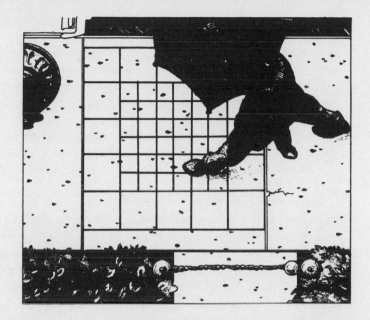

If we imagined laying a grid over the nesting ground with each square of the grid big enough for one bird, then given that a bird occupies a given square on the grid, the probability is very low that you will find another bird in an adjacent square. Birds repel each other, while raindrops are indifferent to one another's presence. Patterns, even if they are random, can have properties that yield important information.

Imagine that the external galaxies are just particles scattered about the space of the universe. Examining this grand view, we might ask: Are the galaxies scattered at random like raindrops or is there a structure to the pattern?

Questions about cosmic organization were first asked by astronomers back in the 1920s when it became clear that the galaxies are vast external star systems thousands of light-years in diameter, similar to our own Milky Way. Astronomers saw that galaxies tend to cluster, as in the Coma or Virgo cluster, and even form superclusters—the first intimation that there was some organizational structure among galaxies. In the early 1950s, the astronomer Gérard de Vaucouleurs first studied the local concentration of galaxies, what he called the "local supercluster," of which the Milky Way is a member. From the examination of the clustering of galaxies grew a new field of inquiry —the study of the large-scale structure of the universe.

In the 1950s, Donald Shane and Carl Wirtanen at the Lick Observatory in California began surveying more than a million galaxies— an almost indigestible amount of data. P. James E. Peebles and his collaborators at Princeton University took the Shane-Wirtanen data and processed them into a two-dimensional map of the sky which became the popular poster "One Million Galaxies." It gives us a fair picture of the distribution of galaxies—a bird's-eye view of the cosmos.

Looking at this picture one can see dense clusters of galaxies forming knots, and between the knots are voids. Some people see evidence of a thin filamentary network structure. Provided the sampling technique is accurate, it seems clear from such maps that there definitely is some kind of large-scale structure to the distribution of galaxies on the scale of tens of millions of light-years. They are not scattered like raindrops. But on the very largest distance scales of billions of light-years—what one would see if one looked at the map through frosted glasses—the universe appears to be a smooth, structureless place. The only structural question at those distance scales is

What is the overall geometry of space—is it open or closed?—a question we address in the next chapter on cosmology.

Astronomers, even in the face of more extensive data than they had in the 1950s, continue to debate issues related to the large-scale structure of the universe. The two extreme positions in this debate are occupied by what we will call the "hierarchists" and the "filamentarians." The hierarchists see a continuous "hierarchical clustering" of galaxies from relatively small distances of several galactic diameters out to the largest distances. Small groups of galaxies combine to make up larger groups and these large groups combine to form still larger groups—a continuous hierarchy of clusters. By contrast, the "filamentarians" see strong evidence for cosmic filaments consisting of thick strings of galaxies on the supercluster scale with large voids between them. The distinction between these two extreme positions is not always a completely sharp one, since deviations from smooth clustering might look like filaments and thick, diffuse filaments look like part of a smooth hierarchical set of clusters.

Recent data analysis seems to support the filamentarians, and most astronomers take this view. Yet others maintain that the filaments appear because of biased sampling methods—one picks the galaxies that support one's viewpoint. It is a bit like reading tea leaves—one looks at a random pattern and sees what one wants. One need only recall the turn-of-the-century debate about canals on Mars—also linear, filamentary structures—to appreciate the problem. In the absence of detailed information, the eye tends to fill in imagined lines between the spots which are in fact not there.

The emphasis of the hierarchical-clustering view is on the smoothness of the clustering transition from small cosmic scales to the very largest—one finds clusters of all sizes. One might compare it to the distribution of people at a very large cocktail party. While a few people may stand alone, most gather into groups or clusters of varying sizes to engage in conversation or listen. Some groups may have just a few people, others many more. Some groups may merge. The people are clearly not distributed purely at random, which is what you might expect if hundreds of tennis balls were thrown around the floor of an empty room. Tennis balls do not cluster together the way people do.

This view of the distribution of galaxies is disputed by the filamentarians, who think there is even more structure to the distribution of galaxies and that the clustering transition is not smoothly hierarchical. They see evidence for a filamentary structure, a network of cosmic strings or surfaces in space around which the galaxies tend to congregate. In other regions of space, between the filaments, there must be great voids. If we went to the cocktail party and drew a wide network of random curved lines on the floor, and then told people that they had to stand on or near a line, the resulting distribution of people on the floor would form filaments; this would be closer to how the filamentarians view the distribution of galaxies.

What do the data say? New data, compiled since the Shane-Wirtanen million-galaxy survey, lend support to the filamentarians. The Estonian astronomer Jan Einasto back in the 1970s began to look at three- (rather than just two-) dimensional maps of the galaxies. This, of course, required determining not only the position of individual galaxies in the sky but also their distance away from us, derived by measurement of their red shift and application of Hubble's Law. Einasto claimed to find evidence for filaments and voids in the small sectors of the sky he examined. Then in 1981 astronomers using similar sampling techniques found a large "hole in space," a region with a 200-million-light-year diameter centered about a point 500 million light-years beyond the constellation of Boötes in our galaxy. This finding seemed to support the filamentarians, who argued that between the filaments there had to be large voids.

A red-shift survey of 2,400 galaxies within a 400-million-light-year radius from us was completed by Marc Davis and his collaborators at the Harvard-Smithsonian Center for Astrophysics in 1981. A local high school student made a three-dimensional display of the Davis data, using pith balls hanging on strings to represent the galaxies. In this display one clearly sees evidence of voids and filaments. Davis remarks, "I think we now have sufficient data to convince everybody that there is a loose, filamentary structure." Theorists like Edwin Turner doing detailed statistical analysis of the distribution of galaxies would concur. He finds evidence for cosmic strings and sheets. It is fair to say that most astronomers are convinced of the existence of the filaments and surfaces. Yet others find this evidence uncompelling, claiming that these patterns are only in the eye of the beholder and that the statistical analysis and sampling are biased. Davis himself admits that the size of the grid used in the computer processing of the data can alter the conclusions.

Why are astronomers so intensely interested in establishing the large-scale structure of the universe—an interest that goes beyond the desire to know such structure merely for its own sake? They know that the large-scale structure of the universe provides a deep clue to its nature. It bears on two related, important issues: the existence of dark matter and the origin and evolution of galaxies. The stakes are high for a correct reading of the cosmic tea leaves.

Astronomers and physicists realize that it is unlikely that gravitational attraction between individual galaxies acting over a long period of time could alone have produced the observed distribution of galaxies. Something else has to be present. A likely candidate is "missing mass" in the form of dark matter—matter hidden within or among the galaxies that cannot be seen by telescopes.

Indirect but good evidence for such dark matter was found as early as the 1930s in studies of clusters of galaxies by the astronomers Fritz Zwicky and Sinclair Smith of Cal Tech. They concluded that the galaxies in a cluster were moving too fast for the cluster to remain bound. It should fly apart, because the mutual gravitational attraction of all the galaxies was insufficient to bind such a rapidly moving system. This was the first good evidence that clusters of galaxies must have some kind of dark matter which provides the additional mass to bind them together. Subsequent studies of other galactic clusters revealed the same characteristics.

More recently there has been accumulating evidence for dark matter even on the scale of superclusters of galaxies. In 1982, R. Brent Tully of the University of Hawaii and J. Richard Fisher of the National Radio Astronomy Observatory in Green Bank, West Virginia, completed a nine-year survey in which they measured the positions and light red shifts of about 2,200 galaxies in the Virgo supercluster. At the center of this supercluster of galaxies sits the Virgo cluster itself, which contains about 20 percent of the galaxies in the supercluster. A book held at arm's length would just cover it in the night sky. At the center of the Virgo cluster sits M 87, a giant elliptical galaxy, with a gravitational maelstrom of other galaxies, stars and hot gas emitting X rays moving around it. Tully and Fisher found that surrounding the Virgo center is an immense disk of galaxies approximately 35 million light-years in diameter and about 6 million light-years thick. Around the disk are streaming cigar-shaped clouds of galaxies directed toward the core—an impressive, somewhat floral configuration of thousands of galaxies. The disk they observe cannot be bound together by just the gravity of the galaxies in it; the galaxies are moving too fast for that. Again dark matter, distributed in the

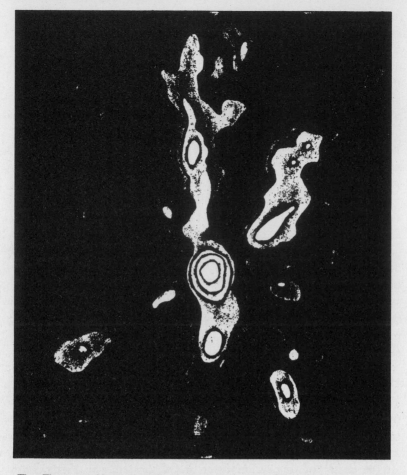

The Flower in Virgo. A radio survey of about 2,200 galaxies in the Virgo supercluster revealed this composite picture. The center of the supercluster is the Virgo cluster, and at its center is the giant elliptical galaxy M 87. Surrounding the central cluster of galaxies are a disk-shaped distribution of galaxies some 35 million light-years in diameter and several "clouds" of galaxies directed toward the center. Similar clusters of galaxies are found throughout the universe.

shape of a disk on the scale of superclusters, could be the responsible agent.

Not only is dark matter present on the scale of distances the size of clusters and superclusters: it is also possibly the main material component of single galaxies. As I described in a previous chapter, over the last decade the velocity of neutral hydrogen and isolated stars orbiting far beyond the visible boundary of spiral galaxies has been measured. If all the mass of the galaxy were in the stars—so that the mass density would fall to zero at the boundary of the visible galaxy—then the velocity of orbiting gas or stars beyond the boundary ought to fall to zero with increasing distance away from the galaxy. Instead, the velocity of the orbiting gas remains constant or even increases beyond the galaxy's edge. This suggests that there is dark matter hidden in the galaxy, and that it reaches out far beyond its visible component. Possibly the visible part of a galaxy—all the stars and gas—is but a mere 10 percent of its total matter; the rest is in a gigantic invisible halo which extends far out in space—so far that it may merge with similar halos from neighboring galaxies.

What *is* the dark matter? Astronomers cannot give a simple answer. Yet they can give reasons why some of the obvious candidates do not qualify. They have considered the possibility that it might be gas, dust grains, frozen hydrogen, collapsed stars like black holes, Jupiter-sized planets. All these possibilities, for varying reasons, have been ruled out. Except possibly for brown dwarfs, nothing in the usual repertoire of matter seems to work.

Theoretical physicists, never at a loss for an answer, have conjectured that electrically neutral quantum particles created in the big bang, such as neutrinos, gravitinos, photinos and axions, make up the dark matter. This idea, at first considered farfetched, has gained support in the last few years, since it seems to shed light on the problem of the origin of galaxies as well. Remarkably, the largest things we see—galaxies and clusters of galaxies—may not only be structured by the smallest things we know—quantum particles—but actually owe their existence to them.

Such dark matter in the form of a fluid of invisible quantum particles may pervade the cosmos, providing 90 percent of its total mass. Galaxies would then constitute but a small fraction of the total mass density of the universe; they would just be going for a ride on the dark fluid. The dark quantum-particle matter, if it exists, may gravitationally congregate in gigantic lumps the size of galaxies or clusters, the size depending on the properties of the quantum particles themselves. Then the galaxies would be like luminous particles which through their distribution reveal the lumps, filaments and clusters of the cosmic fluid of dark matter. There are those who do not share this view and think there may be no relation between the distribution of dark and visible matter.

As astronomers search for answers to these questions, they are also trying to solve the riddle of the origin and evolution of galaxies. The large-scale structure of the universe revealed by the distribution and motion of the galaxies is presumably a remnant of the big bang. Therefore let us go back in time to the big bang and ask, How did galaxies originate out of the primordial fireball?

This is not an easy question to answer. Scientists do know that the original fireball was very homogeneous and smooth, because the microwave background radiation—also a remnant of the big bang and the oldest thing we can see—is distributed the same in all directions in the heavens (once one eliminates the effect of the motion of the earth relative to the background radiation) to better than 1 part in 10,000. Yet there had to be small deviations from this large-scale homogeneity, little "lumps" that developed into galaxies. Looking at the universe this way is like looking at the surface of a lake which on the whole is smooth, but up close has ripples and waves. But why is the universe lumpy?

We are really asking two questions, and it is best to separate them. The first question is What was the origin of the tiny mass-density fluctuations—the deviations from homogeneity—that eventually developed into galaxies? The second question is Given that those tiny fluctuations in the big-bang fireball existed to begin with, how then did such fluctuations grow and develop into the first galaxies? Let us examine these questions one at a time.

Physicists working on the big bang think that the initial tiny density fluctuations started as quantum fluctuations—a necessary part of any physical system, including the gas of quantum particles that was the fireball. There is no way you can avoid them. The only question is How is it that the quantum fluctuations were of the right size—big enough to eventually turn into galaxies but not so big that they would destroy the overall homogeneity of the universe? To answer that question, physicists must examine the universe mathematically at the time when those fluctuations originated, before the first billion-billionth of a second, a period of time for which they have the least confidence that they know what is going on.

Some hold that the structure of the distribution of galaxies we see today is a consequence of properties inherent in those initial quantum fluctuations, the way an oak tree is a consequence of the acorn from which it grew. Hence, the large-scale structure of the universe provides a kind of window on those very early times. Others, contesting this view, maintain with equal confidence—which is not great in either case—that the large-scale structure of the universe is a consequence of the later growth of those fluctuations and is rather independent of the initial specific properties of the fluctuations. Such disputes are part of the forefront of current discussions.

Suppose that physicists could account for the origin of the initial mass-density fluctuations. Then they could show how each fluctuation grows because such a "lump" has only two simple forces acting upon it. First is the energy of motion of the gas particles, proportional to the temperature of the gas, which is trying to disperse the lump. The second force is the mutual gravitational attraction of all the particles, which is trying to contract the lump. Knowing the temperature and mass of the particles in the primordial gas, physicists can calculate the growth and eventual size of the mass-density lumps. They find that with the right temperature and mass conditions, more and more surrounding matter was swept into these lumps by their gravitational pull and they expanded in size as the big-bang explosion proceeded. After about 100,000 or a million years the big bang was over and lumps of gaseous matter populated the universe. True galaxy formation then proceeded.

The scenario for galaxy formation depends crucially on the size and mass of the lumps that populated the universe at this early time. Some scientists think that the first lumps had the mass and size of today's clusters and superclusters of galaxies—very large lumps. Subsequently these big cluster-sized lumps fragmented into galaxy-sized lumps.

Others think that the first lumps were the size and mass of today's galaxies to begin with—relatively small lumps that were the embryos of those galaxies. In this picture, after the galaxies were established they then began moving together to form the clusters and superclusters we see today. Astrophysicist P. James E. Peebles, defending this viewpoint, argues, "If the clusters formed first, then why aren't all galaxies stuck in clusters? Most galaxies are like ours—on the outskirts, in loose groupings, just starting to fall in." The difference between these two views might be summarized as the difference between "first clusters, then galaxies" and "first galaxies, then clusters." Let us examine these alternative views in turn.

The view that clusters and superclusters formed first, as gigantic lumps of matter which subsequently fragmented into galaxies, originated as the "pancake" model of Yakob B. Zel'dovich working at the Institute of Applied Mathematics in Moscow. In the early 1970s, Zel'dovich and his coworkers showed that if there was an initial lump of gaseous matter, its collapse under its own gravity would not be spherically symmetric, the way a balloon deflates. Instead it is more likely that the lump collapsed into a flat sheet—a pancake of matter. At the time the pancake model was proposed, however, there were no known quantum particles that had the right properties to form the pancakes.

This pancake model was revitalized in 1980 when physicists proposed that neutrinos—tiny quantum particles—have a small mass and were a viable candidate for the dark matter. Zel'dovich argued that density fluctuations in a gas of massive neutrinos pervading the universe would grow to form supercluster-sized lumps in the primordial fireball and collapse into neutrino pancakes. Later, hydrogen atoms, formed about 300,000 years after the big bang, would be gravitationally attracted to the preexisting neutrino pancakes, thus forming a dense hot gas. This protocluster of hydrogen gas after a long time fragments into galaxy-sized lumps through a variety of complicated physical processes. Galaxies, in this "pancake" model, are thus relative newcomers to the cosmic stage.

That conclusion turns out to be a problem for the pancake model. Galaxies are about the age of the universe itself—they are old, not young. A further problem is that recent quantum-particle experiments have indicated that at least one neutrino does not have enough mass to pull off the pancake-lumping scenario. These problems have dampened enthusiasm for the pancake model.

The alternative to the pancake protocluster model is the view that galaxies formed first and only recently began to assemble into larger clusters. If the dark matter forms into the smaller galaxy-sized lumps, then it must consist of quantum particles which have not yet even been detected. It is unlikely that these conjectured particles—candidates from particle theory are called gravitinos, photinos or axions—ever will be directly detected in terrestrial experiments because they interact so very weakly with other matter. The only way they can be "seen" is as the conjectured dark matter in galaxies. That seems to many people to be a cheat—inventing a particle to do the job of galaxy formation which is otherwise undetectable.

1 billion years

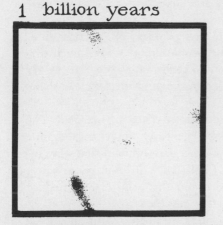

3 billion years

5 billion years

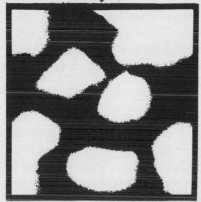

9 billion years

These computer simulations of pancake formation illustrate one theory of the evolution of galaxies. Dark matter (possibly massive neutrinos, if they exist) gravitationally collapses into lumps with the mass of superclusters of galaxies. The collapse is not spherically symmetrical but, instead forms sheets, or massive pancakes, which later fragment into galaxies. This model represents but one attempt to account for the presently observed distribution of galaxies.

But evidently there *is* dark matter in the halos of galaxies, and it could well be a gas of these conjectured quantum particles. These particles could form lumps of dark matter in the early universe of just the right size to subsequently make galaxies. The lumps of galaxy-size dark matter later gravitationally congregate—taking the galaxies with them—to evolve into clusters and superclusters, creating filaments and holes.

Which viewpoint—"clusters first, then galaxies" or "galaxies first, then clusters"—is right? Reasonable people disagree. But such disagreement is a sign of progress, for as Peebles remarked, "It is only recently that the subject has advanced to the point that we can make out positions that seem defensible."

To resolve this issue more observational data are needed. The Space Telescope will begin to probe the early universe, the time before the era of quasars, and astronomers are hopeful that new information on the early evolution of galaxies will be provided. Perhaps the puzzle of whether galaxies and then clusters formed or clusters and then galaxies formed can be resolved.

Yet no matter how far back in time one probes with an optical telescope, and no matter how powerful the instrument is, one cannot reach back before the first 300,000 years. The universe is opaque before that time. One cannot hope to see the initial fluctuations in the primordial fireball that eventually give rise to galaxies. They are shielded from view, hidden in the big bang. But their structure may be the clue to all subsequent galactic evolution and the large-scale order of the universe.

We do not know why the universe is lumpy—a slight lumpiness, to be sure, but very important. The galaxies, stars, planets and humanity itself are part of that lumpiness. Theoretical physicists, guided by experiment and observation, have extrapolated current quantum theories into energies which are comparable to those before the first microseconds of the big bang. Here they speculate about those initial quantum seeds from which the galaxies grew—speculations on the edge of current research. I will report on their work in a subsequent chapter.

Fritz Zwicky, who helped discover the great clusters of galaxies, once remarked that they were "the last stepping-stone to the study of the universe as a whole." The study of the whole universe is called cosmology, and represents the end point of human thinking about the large-scale order of space and time. To this subject we now turn.

7

Classical Cosmology

The fabric of the world has its center everywhere and its circumference nowhere.
          —Cardinal Nicolas of Cusa, fifteenth century

Ludwig Wittgenstein, the philosopher, in his *Lectures on Ethics* describes a peculiar experience of his by saying, "I believe the best way of describing it is to say that when I have it *I wonder at the existence of the world.* And I am then inclined to use such phrases as 'how extraordinary that anything should exist' or 'how extraordinary that the world should exist.' " Martin Heidegger, the existential philosopher, sees such experiences as reflecting the root question of metaphysics: "Why is there any Being at all—why not far rather Nothing?" This question articulates the puzzle posed by the fact of existence: who needs the complexity of the world in the face of the simplicity of absolute nothingness?

Reflecting on this question, I find that it provokes ambivalent emotions. I feel anxiety as I imagine the unfathomable, silent abyss of nothingness which might have been, and I experience open wonder as I acknowledge the mysterious yet simple fact that the universe exists. I do not know the answer to the philosopher's question, nor am I convinced that it can be simply answered. But it is the kind of question that rubs our noses into the weirdness of reality, takes us to the threshold of madness and tells us to "wake up." The strangeness of concrete existence is a feeling that never quite leaves us no matter how familiar or comfortable we become with the world. Nor should it. That odd feeling provoked by our recognition of the reality of existence rejuvenates our wonder and curiosity about the universe, restoring the child in each of us. And that wonder is the beginning of science.

As a child I often looked up at the night sky filled with stars, but I did not see what astronomers tell us is there. What I saw was a heavenly vault, a gigantic black spherical lid covering the surface of the earth. In this opaque lid there were holes which revealed the celestial fire on the other side—holes that appeared as stars. Slowly, as the evening progressed, the vault with its brilliant flickering orifices moved across the sky. I could almost see it move. On bright moonlit nights I thought I saw moonlight reflected off the black vault like light from a movie screen reflected from the walls of a theater. The sun and moon were held in place by other transparent spheres which carried them through the sky.

Such was my child's theory of the universe. I do not recall anyone's telling me about this view of the heavens, but later upon reading Aristotle, the ancient Greek philosopher, and Ptolemy, the Alexandrian astronomer, I recognized my cosmology in theirs. Aristotle saw the universe as a system of fifty-six celestial spheres carrying the heavenly bodies. Ptolemy elaborated on this system, correlating it with quantitative observations. The Ptolemaic cosmology, in spite of its complicated cycles and epicycles, accorded with common sense so completely that it dominated Western cosmology for a thousand years, until the time of Copernicus. And this is no accident of ignorance. We forget today that it is Aristotle's physics and Ptolemy's cosmology which correspond to common sense, not Newton's physics and Copernicus' cosmology, which are already great abstractions from our ordinary experience.

The world "theory" derives from the Greek "to view"—a theory is a picture of reality. My childhood theory of the heavens was an example. Though there are no celestial spheres as I had thought, this "theory" brought a satisfying coherence to my experience which was not part of mere perception. That is the great power of theory as a picture of reality—it orders our experience in new ways and renders the complexity of our perceptions intelligible. But that power of theory is abused if we confuse our picture of reality with reality itself, confuse the map with the territory. We should never forget—as we almost always do in practice—that our theories are only maps of reality. All the theories coming from natural science, in spite of their coherence, predictive power and depth of insight, are nonetheless only ways of describing material reality and not reality itself. For example, according to Newton's theory the laws of motion are differential equations. But the planets, as they move in their orbits about the sun, are no more solving differential equations which tell them how to move than they are attached to celestial spheres. They are simply moving. It is we who invent the theory, solve the differential equations and see whether our picture corresponds to reality. The planets have no such problem.

Physical theories and cosmologies can be thought of as maps of reality. But this does not imply that such maps are arbitrary inventions. Some maps are far better and more accurate than others, encompassing more of the territory of reality. We can compare different maps and see which works best. For example, my childhood Aristotelian "map" of the heavens fails once we begin to examine the universe more closely. The Aristotelian map is replaced by a Newtonian one, and this in turn is replaced by an Einsteinian one. As scientists learn more about reality through observation and experiment the maps change to accommodate the new discoveries. Sometimes the experimental discoveries are so perplexing that there is no way to describe the new territory on any existing map. Then the very rules for making maps have to be changed, as was done at the beginning of this century with the advent of relativity and quantum theory. But we keep on making new maps, and reality just goes on existing. It is important to bear this in mind as we discuss cosmology. All cosmologies are models of the universe, not the universe itself.

According to the distinguished English cosmologist Herman Bondi, "Cosmology is the field of thought that deals with the structure and history of the universe as a whole." Yet how can we deal with the whole universe, which, by definition, includes every physical thing? While we can look at distant galaxies from the outside, we cannot look at the universe from the outside, because there is no outside to the universe. It makes sense to talk about the relative locations of stars and galaxies in space, but it is meaningless to talk about the location of the universe in space. Is the space of the universe infinite or finite? If it is finite, where is the edge? Did the universe have an origin in time, and if so, what existed before the universe? How will the universe end? Clearly, in considering the whole universe in space and time we are considering an entity of a completely different kind and not just another, larger astronomical object. Consequently, new and perhaps unfamiliar concepts must be invoked if we are to provide a framework for our thinking and attempt to answer the questions we have raised.

Reflective people have thought about such questions for centuries. Yet, remarkably, only in the last few decades has cosmology, once the province of merely speculative thought, become an empirical science. Various cosmological models which attempted to answer cosmic questions are now subject to observational test. For example, the old steady-state model of the universe, which maintained that the universe is infinitely old and continues to look much the same over time, can now be ruled out (or at least rendered extremely implausible) by astronomical observations. The capacity to rule out well-defined models is the mark of a mature empirical science.

Two major advances transformed cosmology from a speculative to an empirical science. The first advance, on the theoretical side, was Einstein's creation of the general theory of relativity, a comprehensive theory of space, time and matter which provided a new conceptual framework for our thinking about the universe as a whole. The second advance that brought cosmology to its modern form was the deployment of the powerful new astronomical instruments—the big reflectors and radio telescopes. Einstein's theory does not single out a specific cosmology or structure for the universe; it provides the framework, not the details. To decide what is the actual structure of the whole universe in space and time requires, as always, detailed observation, and for this new instruments were needed.

For the first decades of this century as astronomers looked deeper into space they continued to see a hierarchy of larger and larger structures: from stars to galaxies to clusters of galaxies, all expanding with the universe. But for the last several decades astronomers have been exploring the large-scale structure of the universe itself, and they have found that this hierarchical structure of bigger and bigger lumps stops. On the very largest distance scales of hundreds of millions of light-years one begins to see the smoothness of the universe. This smoothness seems to be a global texture of the cosmos, not just some local property of our region of space. For the first time we are seeing spatial features of the universe as a whole. The study of that large-scale, smooth, homogeneous space, its development in time and how it influences the matter within it is the true scientific province of the contemporary cosmologist.

Let us now examine these two main threads of cosmology—the theoretical and the observational—in more detail, beginning with the modern theory of space and time.

For the moment forget about actual physical space and try to imagine pure, empty three-dimensional space. Suppose you are in a rocket ship, sitting still in deep space, and before you turn on the rocket engine you leave a laser device floating in space to mark your place. It emits a light beam, which you follow faithfully, never turning back. After a while you see the same laser appear up ahead. It seems as if you had gone in a circle. Next you try a different direction, but the same thing happens. This space is clearly not an ordinary one in which, if you take off, you never return. It is an example of a non-Euclidean space, and while it may seem strange, it is mathematically possible.

The mathematical description of such "curved" spaces was first given in complete generality by the great nineteenth-century German mathematician Bernhard Riemann. Usually we think of physical, empty space as flat, so that if we were to use laser beams to form the sides of triangles, cubes and other geometrical figures they would obey the theorems of Euclidean geometry. If we took off in a rocket ship in a straight line, we would never return. But Riemann's work generalized the idea of space to include the possibility of non-Euclidean geometry as well, in which the space is not flat but possesses a curvature. It would be like generalizing two-dimensional spaces to include not only the flat space of a piece of paper but also curved surfaces like that of a pear. Riemann showed how the geometrical curvature of non-Euclidean space was completely described by a mathematical object called the curvature tensor. In principle, by using laser beams in three-dimensional space and measuring angles and distances with them, we can determine Riemann's curvature tensor at each point in that space.

The beauty and power of Riemann's geometrical work are that it set on a firm mathematical foundation the description of arbitrarily complicated curved spaces in any number of spatial dimensions. We can easily visualize most curved two-dimensional spaces like the surface of a sphere or a doughnut. Our ability to visualize curved three-dimensional space breaks down. Yet Riemannian mathematical methods show us how to deal with such spaces—mathematics can guide us even where the visual imagination breaks down.

The mathematics of curved spaces and their associated non-Euclidean geometries was rather well understood by mathematicians in the first decades of the twentieth century. But these advances appeared to be in the realm of "pure" mathematics, and the full impact of these ideas for the real physical world came only with Albert Einstein's invention of the general theory of relativity in 1915–1916.

The general theory of relativity grew out of Einstein's 1905 special theory of relativity, a theory that established a new kinematics for physics: the laws of transformation which relate space and time measurements made by one observer to those of another moving at a constant velocity relative to the first—so-called "inertial observers." Einstein's kinematics approximated the Newtonian kinematic laws for small relative velocities, but differed markedly when the relative velocity of the two observers was near that of light. In Newtonian kinematics, the laws of transformation for space and time measurements are separate; the time transformation is independent of the position and relative velocity of two inertial observers. Thus time can be taken to have a universal, "absolute" significance for all inertial observers. But in Einstein's kinematics, space and time measurements become intimately interrelated; time is not absolute but is relative to a particular inertial observer.

Einstein, reasoning from two general postulates—that absolute uniform motion of an inertial observer is undetectable and that the speed of light is an absolute constant—correctly deduced the new laws of space-time transformations. But it was Hermann Minkowski, a mathematician, who pointed out their geometrical interpretation. Minkowski showed that if one did not view the three space dimensions and the one time dimension as separate entities, but instead joined them together into a four-dimensional space-time, then Einstein's new transformations could simply be seen to correspond to rotations performed in this four-dimensional space-time. This was an enormous simplification, creating a new perspective on space and time. As Minkowski commented in 1908, "Henceforth space by itself and time by itself are doomed to fade away into mere shadows, and only a kind of union of the two will preserve an independent reality."

Einstein at first did not care for Minkowski's interpretation, thinking it to be of only formal, mathematical significance and not of real physical significance. Yet he subsequently changed his mind, adopting the four-dimensional viewpoint fully when he turned to inventing general relativity.

Einstein, attempting to extend the principle of his earlier theory of special relativity to include gravity, already intuited that to generalize the relativity principle required the consideration of non-Euclidean geometry. With the aid of his friend and former classmate Marcel Grossmann, a mathematician, Einstein studied Riemannian geometry, and here he learned the proper mathematical language for expressing his physical intuitions about gravity and geometry. After more than a decade of struggle and frustration, which would have stopped a lesser man, Einstein finally succeeded in obtaining a set of equations—the Einstein equations—that specified the curvature of four-dimensional space-time in terms of the matter present in that space. Einstein's equations emphasized concepts such as invariance, symmetry and geometry which had a profound impact on the future development of physics. Gravity was reduced to geometry. Matter, space and time became unified. Non-Euclidean space was not just an intellectual curiosity but a correct description of reality.

Einstein's equations, which predicted the curvature of space-time near the sun, were supported by observations a few years after he wrote them down. But recently they have been experimentally tested with greater precision through the use of artificial satellites and radar ranging methods, and were beautifully confirmed. But as Einstein realized even as he was inventing the theory, his equations describe not only the curvature of space-time in the solar neighborhood but also the curvature of the space-time of the entire universe. The Einstein equations provide the conceptual framework for modern cosmology.

Einstein was the first to apply his equations to the problems of cosmology. Like most physicists of his time, Einstein believed that the universe was static and unchanging (this was before the discovery of the expansion of the universe), and so he assumed this when he looked for solutions to his equations. He then solved his equations and reached the seemingly absurd conclusion that the universe changed. For this and other reasons, Einstein decided to alter his equations by adding a new term—the "cosmological term." With this modification, Einstein, in 1917, found a static, curved space, filled with a uniform gas of pressureless matter, that was the solution he sought. The universe did not move. Had he stuck with his original equations without the "cosmological term" and looked for solutions that did change in time, Einstein could have anticipated the discovery of the dynamic, expanding universe.

In the same year, 1917, the Dutch astronomer Willem de Sitter found yet another solution to Einstein's equations with the cosmological term added but with no matter—an empty universe. De Sitter's solution to the Einstein equations could be interpreted as an expanding space like the stretching surface of a ball of rubber.

So there were two cosmological models based on Einstein's equations: Einstein's cosmology, which had a static space filled with matter, and De Sitter's cosmology, which had an expanding space devoid of matter. De Sitter's solution, since it represented an empty yet moving universe, seemed to many people quite absurd. A common opinion of the time was that Einstein's equations had not shed much light on cosmological problems.

Then, in 1922–1924, a Soviet meteorologist who became a professor of mathematics at the University of Leningrad, Alexander Friedmann, found the dynamic solutions to Einstein's original equations (without the cosmological term) which we now believe correctly describe cosmology. Friedmann's work, unfortunately, was entirely ignored, even though his papers were published in a leading scientific journal and Einstein was aware of them. Not until Georges Lemaître, "the father of the big bang," independently rediscovered his equations in 1927 did cosmology gain its modern framework. And Lemaître's work was also ignored until the prestigious astronomer Arthur Eddington pointed out its importance in 1930.

Subsequent contributions were of a mathematical nature. Howard P. Robertson and Arthur Walker, two mathematicians, showed that Friedmann's solutions were actually the most general solutions to Einstein's equations, provided one assumed the universe was spatially homogeneous and isotropic. They further demonstrated that in this case four-dimensional space-time could be separated into a curved three-dimensional space and a single time common to all "co-

---

OPPOSITE:
Two-dimensional analogues of the three possible homogeneous and isotropic three-dimensional FRW spaces. Here, a two-dimensional scientist explores these spaces. On the top is the infinite flat space, in which parallel laser beams never meet and the sum of the angles in a triangle is 180 degrees. In the middle is the finite space of positive constant curvature, for which parallel laser beams converge and meet and the sum of the angles in the triangle is more than 180 degrees. On the bottom is the infinite hyperbolic space of constant negative curvature, in which parallel laser beams diverge and the sum of the angles is less than 180 degrees. The two-dimensional space of constant negative curvature (unlike the other two spaces) cannot be embedded in a three-

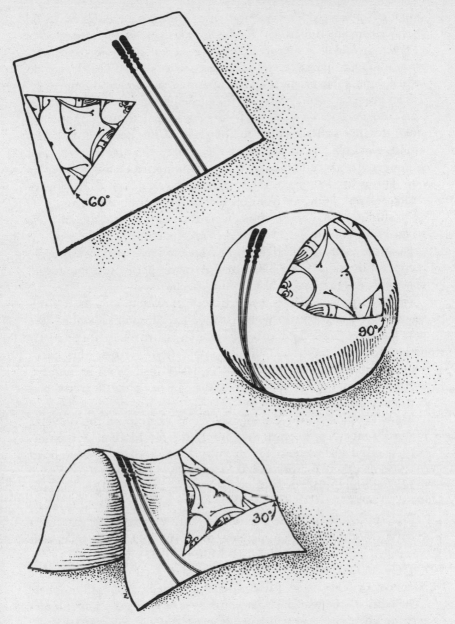

dimensional space, as has been attempted in this illustration. For that reason, it does not appear to be homogeneous and isotropic and has a special place —the saddle point.

moving" observers. Today, modern cosmological models based on these Friedmann-Robertson-Walker solutions are often referred to as "FRW" cosmologies. Before examining them in detail, let us see why from a modern point of view they are so attractive. That has to do with a remarkable property of the observed universe.

The universe seems to be a complicated structure—with planets, stars and galaxies. Yet all these organizational features appear on small distance scales relative to the scale of the universe itself. As astronomers observe larger and larger scales, the universe becomes smoother and more homogeneous—the lumpiness tends to average out. This is like looking at the surface of the earth, which has lots of structures and "lumps," from on high in a jet plane and seeing it as quite smooth. Such observations lead us to assume that the universe when viewed at the very largest distances is both homogeneous—it appears the same irrespective of one's location in it—and isotropic— it appears the same in all directions. A space that is isotropic for *all* observers, not just one, is also homogeneous.

Further strong evidence supporting these assumptions came from the detection in 1965 of the microwave background radiation—the heat left over from the big bang—which is the oldest thing so far detected in the universe. Within observational errors, this background radiation is completely isotropically distributed around us, indicating that even as far back as the big bang the universe was extremely isotropic.

Many people find the homogeneity and isotropy of the universe rationally satisfying because they imply that no location in the universe is privileged or special in any way. The alternative would be to suppose that there *is* a privileged location, and then one would have to ask why that place was privileged and not some other. But one does not even have to ask such a question if, as Einstein expressed it, "all places in the universe are alike." This attractive idea has been given the distinction of a principle—the cosmological principle, as it was called by cosmologist Edward Milne in 1933.

The cosmological principle harks back to a famous aphorism of Nicholas of Cusa, the fifteenth-century theological philosopher: "The fabric of the world has its center everywhere and its circumference nowhere." Copernicus, in his model of the solar system, removed the earth from the center, so the earth was not a privileged planet. Centuries later, Shapley showed that the sun is not in a privileged place either; we are far from the center of our galaxy. Today we even know that our galaxy has no special location among the millions of galaxies observed. There seems to *be* no "special" place.

But the cosmological principle, as a scientific proposition, could be wrong. For example, the entire universe and all the galaxies in it could be rotating. The universe would then have an axis of rotation, a preferred direction, and would not be isotropic. The cosmological principle and the homogeneity and isotropy of the universe it implies are subject to falsification within the observed universe. But today most evidence favors it.

· The FRW cosmologies incorporate the cosmological principle; that is why they are so attractive. By assuming, in accordance with the cosmological principle, that three-dimensional space is homogeneous and isotropic, Robertson and Walker showed mathematically that there could be only three such geometrical spaces. Not surprisingly, two of these corresponded to the solutions of the Einstein equations that Friedmann had already found. The three spaces were the flat space of zero curvature (which Friedmann had not found), the spherical space of constant positive curvature and the hyperbolic space of constant negative curvature. In the flat space, parallel laser beams never meet; it is an open space of infinite volume. In a spherical space, parallel laser beams converge; it is a closed space with a finite volume. In this space you can fly straight away and return to your starting point. In the hyperbolic space, parallel laser beams will diverge; the space is open and has infinite volume.

Analyzing these spaces through the use of Einstein's equations, one finds that the curvature changes in time. In the flat space, for which the spatial curvature is zero, the relative scale of space and time measurements changes. From these dynamical solutions to Einstein's equations one can only conclude that the universe cannot be stable— it must change by expanding or contracting—and the space of the universe is stretching.

These solutions thus anticipated Hubble's Law, which implies the expansion of the universe. For if galaxies are placed in an expanding space, then like markers put in the space they too will move apart— the so-called "Hubble flow." Hubble's discovery thus lent powerful observational support to the FRW cosmologies and the dynamic universe.

Yet it did not answer the further question of which of three possible spaces—flat, spherical or hyperbolic—we are living in. Answering that question turns out to be very difficult. But the fate of the universe hinges on the answer, because the flat and hyperbolic geo-metries can correspond to open universes and continue to expand forever, while the spherical, closed universe eventually stops expand-ing and recontracts; it has a finite lifetime. A variety of observational methods have been proposed to try to answer the question; none are very successful. I mention two of them.

The first method consists of trying to observe and measure the rate of slowing of the expansion of the universe. Hubble's data were consistent with a constant rate of expansion; but each of the three FRW cosmologies implies that, although it is occurring very slowly, the expansion of the universe is decelerating. By measuring that de-celeration precisely, one could determine which of the three specific FRW geometries applies to the real world and find out whether the universe expands forever or eventually collapses. Measuring the de-celeration—the variation from a constant expansion rate in time—requires careful observation of the most distant galaxies and checking if there is any variation in Hubble's constant of proportionality be-tween the recession velocity of the distant galaxies (measured from the galactic red shift) and distance (measured by some other means).

Unfortunately, distance estimates to faraway galaxies are ex-tremely difficult to make. In part this difficulty is due to the fact that galaxies are probably evolving, altering their luminosity in unknown ways so that "standard candles" to gauge distances by luminosity rements become unreliable. Consequently one cannot determine the variation in "Hubble's constant" over time, nor find out the rate of deceleration that would reveal whether the universe is open or closed.

A second method of determining whether the universe is bound to collapse consists of first establishing the average matter density of the universe—the density of matter in the universe if we smeared it all out uniformly. According to the FRW cosmologies there is a crucial parameter called Ω, the ratio of the observed average matter density in the universe to the "critical density" of about 10^{-29} grams per cubic centimeter—approximately ten hydrogen atoms per cubic meter. Hence if we but knew the average matter density, we would also know the value of Ω. If Ω turns out to be less than one, then the geometry must be hyperbolic; if equal to one, flat; if greater than one, spherical. Knowing the value of Ω would settle the question whether or not the universe will collapse.

The problem of determining the average matter density is that matter in the universe can consist of both visible matter, like stars and galaxies, and invisible, dark matter like black holes or micro-scopic quantum particles. The visible, luminous parts of galaxies give a value of Ω of about one one-hundredth. Astronomers can directly measure the density only of the visible matter. If one assumes that 90 percent of a galaxy's mass is dark matter, this would imply a value of Ω of about one-tenth. And on the largest scale of clusters of gal-axies, they do find a contribution to the parameter Ω of about one-tenth to two-tenths. If this were the whole story, we would conclude that we are living in an open, hyperbolic universe. Unfortunately, no such simple conclusion may be reached because of possible further dark matter. As we have seen, there is good evidence for such matter. In fact, the dominant material component of the universe could very well be dark matter and the visible component, the galaxies and stars, but an insignificant part of the total mass of the universe.

The upshot is that today we have no reliable way of knowing whether the universe is open or closed. This troubles a lot of people who are anxious to know the fate of the universe. But in my opinion, they are troubled by the wrong problem. Astronomers already know that the crucial cosmic parameter Ω is greater than one-tenth and less than two—a range of values rather close to one. Why is this so? The quantity Ω could have any value; it could be fifty or one-thousandth. The real puzzle is why Ω is so close to one; why are we sitting on the edge between open and closed universes? That is the real problem, one we will ponder in subsequent chapters on the very early universe.

Although astronomers are unable to answer detailed questions about the large-scale geometry of the universe, they have made re-markable progress in understanding its development over time. Today most scientists maintain that the universe evolved from a hot, dense gas of quantum particles which subsequently expanded rapidly —an explosion called the "hot big bang." Everything in the universe is a remnant of that explosion. But such a uniform consensus about the evolution of the universe was not always the case.

Not so long ago scientists were divided into those who, following the spirit of the original Einstein–De Sitter cosmology, believed the universe was in a steady state existing from the infinite past to the infinite future and those who, following the spirit of the Friedmann-Lemaître cosmology, believed the universe was extremely different in the past and had a definite origin. One can scarcely imagine two more opposite viewpoints. This opposition was extremely important for the birth of cosmology as an empirical science. Not only did the need to settle the issue promote the search for cosmologically significant data, but the proponents of each position did complex calculations to defend it, calculations which in the end proved to be more valuable than the position they were defending. Let us examine the not-so-subtle dialectic between the "big-bangers" and the "steady-staters."

Georges Lemaître is rightly called "the father of the big bang" because he emphasized as early as the 1930s that the Einstein equations implied that the universe must have begun as a very dense state of matter—a "primeval atom," as he called it. But the modern version of the big bang began when the physicist George Gamow, pursuing Arthur Eddington's challenge to find a hotter place than the center of a star, started research on the early universe. Like Lemaître, he realized that if one went back in time the universe would contract and the matter in it would be squeezed together, getting very hot—hotter than the interior of a star. That implied that atomic nuclei could be synthesized in the big bang just as they are in stars. He and two students, Ralph Alpher and Robert Herman, then calculated how these atomic nuclei could get cooked up in the hot big bang starting from hydrogen, the simplest nucleus.

But their idea that most of the heavier elements were synthesized in the big bang turned out to be wrong (heavy elements are made in the interiors of stars or in a supernova process). However, by taking on this problem of the early universe they also began thinking about heat left over from the big bang. They reasoned that the heat of the big-bang explosion must still exist, because unlike the heat from a fire or a star, it has no place to which it can escape—there is no "outside" to the universe. The heat would be manifested as a low-temperature bath of radiation pervading the entire universe. Furthermore, the temperature could be estimated. Alpher and Herman summarized their prophetic conclusion in 1948: "The temperature in the universe at the present time is found to be about 5°K." This prediction, based on the big-bang theory, should be compared with the statement of A. A. Penzias and R. W. Wilson summarizing the results of their observations made seventeen years later: "Measurements of the effective zenith noise temperature . . . at 4080 MHz have yielded a value about 3.5°K higher than expected. The excess temperature is, within the limits of our observations, isotropic." This direct observation of the big-bang radiation—at least, that is the simplest interpretation—was the clinching evidence for the big-bang theory.

But before such evidence became available many scientists found the steady-state theory very attractive, a theory invented by Herman Bondi, Thomas Gold and Fred Hoyle in 1948. Their basic idea was that as the universe expands, new matter is continuously and spontaneously created in the space opening up between the galaxies. This new matter eventually forms new stars and galaxies. The authors of the model showed that the required continuous creation of matter in the void of space was so small that there was no conflict with any observation. On the basis of this reasoning, they concluded that in spite of the observed expansion of the universe, it could continue to look rather much the same over long periods of time. In the distant past or distant future the average density of galaxies remains the same because new galaxies are being continuously created. The universe, according to this model, is not only uniform in space but also uniform in time; it exists forever the same. With one sweeping hypothesis the problem of the origin of the universe could be solved: it had no origin. A quality of eternal sameness characterizes this cosmology, a sameness expressed by the author of Ecclesiastes: "The thing that hath been, it is that which shall be; and that which is done, is that which shall be done; and there is no new thing under the sun."

In spite of its attractiveness, the steady-state cosmology posed problems. First, it was never clear why the rate of creation of matter was just right to match the expansion; it was a gratuitous assumption. The continuous creation of matter required by the model also violated the ordinary Einstein equations; they had to be modified. These problems might be brushed aside by the adherents of the model. But a more serious problem was Gamow's earlier claim that the heavy elements were all synthesized in the big bang ("big bang" was originally a term of derision coined by Fred Hoyle). According to the steady-staters, there was *no* big bang, so all the elements must get cooked up inside stars. Proving this was a great challenge. So in spite of the fact that the steady-state cosmology seems wrong today, its proponents, in order to defend the theory, did the first correct calculations of the synthesis of heavy elements in stars. They were done first by Hoyle and later in 1957 by William Fowler and Margaret and Geoffrey Burbidge and independently by Alastair Cameron.

The death knell of the steady-state cosmology began to sound in 1961 when Martin Ryle and Peter Scheuer published counts of radio sources. They reported that the density of radio sources increased as one went deeper into space. Since this distant space corresponded to the distant past, their findings clearly contradicted the requirement of the steady-state model that said the universe was unchanging in time. But some people thought the counts were ambiguous, and a controversy broke out between Hoyle and Ryle.

The final blow to steady-state cosmology came in 1965 from the detection of the microwave background radiation, the heat left over from the big bang, by Penzias and Wilson. As far back as 1941 astronomers knew that some molecules in interstellar space were activated, and in 1965 George Field suggested this was due to radiation. Radio engineers were also troubled by a persistent noise in their antennas but did not know its cause. With retrospective insight we now know that these effects were due to the cosmic background radiation.

Penzias and Wilson at Bell Laboratories in Holmdel, New Jersey, had a large horn-shaped antenna hooked up to a radio receiving device cooled down to nearly absolute zero so that it would be very sensitive to low-temperature radiation. Try as they would, they could not eliminate an irreducible noise level in their receiver which corresponded to a background of radiation with a temperature of about 3 Kelvin.

Meanwhile, not far away at Princeton University, Robert H. Dicke, an experimental physicist, had been thinking about an early hot universe and suggested that there was a way of detecting the radiation it left. A similar suggestion had also been made in 1964 by two Russians, Igor Novikov and Andrei Doroshkevich. Dicke prevailed upon two colleagues, P. G. Roll and D. T. Wilkinson, to build a small microwave detector and have a look. Peebles, at Princeton, aware of Dicke's suggestion but unaware of the earlier work of Alpher and Herman, was redoing the calculations of the temperature of the background radiation. Via a roundabout route, Penzias and Wilson found out about Peebles' work and were told by Dicke that they had probably detected the big-bang radiation. The paper of Penzias and Wilson reporting their measurement of the background-radiation temperature was preceded by a paper by Dicke, Peebles, Roll and Wilkinson offering the cosmological interpretation of that temperature. When Penzias and Wilson saw their work reported in newspapers, they realized they had made a major discovery.

Before this discovery, Peebles' experimental colleagues at Princeton had already set up a horn on the roof of a campus building for the express purpose of detecting the radiation. They measured the radiation at a shorter wavelength than did Penzias and Wilson. The additional data they acquired fitted in perfectly with the expected wavelength distribution of the big-bang radiation.

The discovery of quasars also supported the big-bang idea. Maarten Schmidt, who did major observational work on quasars, asserted in the 1970s that there were 1,000 to 10,000 more quasars 2 billion years ago. If quasars were extremely distant, as most astronomers maintained, Schmidt's observation that there were many quasars in the past would be unambiguous evidence in favor of a changing, evolving universe and evidence against the steady-state model.

After these dramatic discoveries the adherents of the steady-state cosmology dropped to a small fraction of all scientists. As Dennis Sciama, once a steady-state proponent, remarked, "Taken together with the evidence from the radio-source counts and the quasar red shifts, the excess background of radiation creates very grave difficulties for the steady-state theory." The big-bang model triumphed; the steady-state model became a museum piece.

Today, with the steady-state cosmology disregarded, people often forget the original motivation for creating the model. First, the logical and conceptual simplicity of the model neatly eliminated the vexing problem of the origin of the universe. Second, in 1948 there was a serious problem with the big-bang idea: the age of the universe estimated from the expansion rate was less than the age of the solar system—an absurd discrepancy. The steady-state model, by contrast, presented no such difficulty. But this was before Walter Baade's 1952 astronomical work which rescaled the size of the universe by a factor of 2, increasing the age of the universe by a comparable amount and thus eliminating the discrepancy. Baade's recalibration of the distance and time scales of the universe made the big bang a possible model.

With the observational evidence coming in so strongly in favor of the big-bang cosmology, theoretical physicists returned with new-found confidence to calculating the heavy elements that would have been produced in the primordial explosion. By then it was clear that most heavy elements are made inside stars by nuclear burning; but some—distinguished by the term "primordial elements"—must have been made in large amounts in the intense heat of the big bang. Of these helium is the most abundant.

The majority of visible matter in the universe is hydrogen, but about 27 percent is helium. All the other elements make up only a few percent. Already in 1964, before the discovery of the background radiation, Peebles at Princeton and Yakob Zel'dovich in the Soviet Union, independently of each other and unaware of the earlier work of Gamow, Alpher and Herman, turned to doing calculations of the helium abundance and estimating the current temperature of the universe. Hoyle (in spite of his opposition to the big-bang idea) and his collaborator R. J. Tayler in the same year showed that stars burning for the entire age of the universe can convert only about 2 or 3 percent of their hydrogen into helium. Consequently, the stars could not have made enough helium to account for what we currently see. They estimated the amount of helium produced in the early stages of the big bang to be 36 percent. Subsequently other physicists, including Hoyle, William Fowler and Robert Wagoner, did further careful calculations of the synthesis of helium from hydrogen in the big bang and derived a figure of about 25 percent—just the right amount to account for what is observed. The results of these calculations depend sensitively upon the details of the big-bang explosion—the rate of expansion of the universe and the properties of interacting quantum particles in the primordial gas—so the fact that the observed amount of helium is correctly predicted must be viewed as a great triumph of the big-bang model.

Yet even more remarkable is the prediction of the relative abundance of deuterium, whose atomic nucleus consists of a single proton and a single neutron. Deuterium cannot be made in stars and survive (though possibly in supernovas). But astronomers observe a tiny amount of deuterium—about one-hundredth to one-thousandth of a percent—in the universe. The only viable explanation is that all the deuterium is primordial—it was made in the big bang. Indeed, the observed amount of deuterium was easily accounted for when physicists did calculations based on this assumption.

Calculations of the relative abundances of light primordial elements like helium, deuterium and also lithium depend on properties of the universe when it is only a few minutes old. The fact that these calculations predict with such precision the properties of the presently observed universe—properties that are otherwise without a coherent explanation—cannot be accidental. For this reason many scientists are confident that they actually understand the state of the universe when it was only a few minutes old. Curiously, they feel they understand the state of the universe better for the period spanning the first few seconds to the first few hundred thousand years than for either earlier or later times. This is because the universe is rather simple during that period—it is essentially a gas of particles whose interactions are known. Complications have not yet set in.

For times longer than a few hundred thousand years, the formation of galaxies begins. Although the laws of physics that describe the individual particle interactions for galaxy formation are understood, the complexity of the physical processes makes it hard to sort out which ones played the dominant role. We understand the contemporary universe of the last few billion years fairly well primarily because observations can guide us in making models of stars and galaxies. There is a "missing link" corresponding to the period from about several hundred thousand years after the big bang, when the first galaxy or star formation was initiated by contracting hydrogen gas, to about 1 or 2 billion years after the big bang, when the galaxies have already formed. New astronomical instruments will provide clues about this mysterious period in the coming years. With a bit of good fortune in getting new data, and with insight and persistence, the origin of galaxies will become as well understood as the origin of stars is today.

For times earlier than the first few milliseconds of the universe, it is not clear that we even know the detailed laws of quantum physics that govern the interactions. It is a realm of time that is at the frontier of current research. Here the problems of cosmology and quantum physics merge into a common problem.

To study the very early universe, cosmologists join forces with quantum-particle theorists. They find that the universe is so intensely hot and the particles so energetic that the quantum properties of matter become extremely important. The very early universe can be thought of as a powerful particle accelerator, a new proving ground for the most advanced models of quantum particles. I will devote the next part of this book to describing this exciting area of research.

But no matter how far back in time we go, there is always something there—a gas of quantum particles. What happens as we approach the origin of the universe itself? How can we even think about that? Where did the matter come from? Did time have any meaning? Did the universe begin in chaos or simplicity? What went on before the universe was created? While the theoretical physicists who speculate about these profound questions have no simple or unique answers, they have begun to approach these questions in a rational, mathematical way. That represents progress over previous attitudes. It is remarkable that the current concepts of quantum-field theory and relativity theory, while probably still quite incomplete, do provide a beginning framework for thinking about the profound and difficult problem of the very origin of the universe. We do not have to surrender to our ignorance, nor give up on the attempt to rationally comprehend nature, nor abandon ourselves to a self-serving illusion.

The attempt to understand the origin of the universe is the greatest challenge confronting the physical sciences. Armed with new concepts, scientists are rising to meet that challenge, although they know that success may be far away. Yet when the origin of the universe is understood, it will open a new vision of reality at the threshold of our imagination, a comprehensive vision that is beautiful, wonderful and filled with the mystery of existence. It will be our intellectual gift to our progeny and our tribute to the scientific heroes who began this great adventure of the human mind, never to see it completed.

William Herschel, who began his career as a musician, became the great eighteenth-century astronomer whose observations led the way to the dynamic and evolving view of the universe we hold today. The discoverer of Uranus, he also showed that distant binary stars move in accord with Newton's laws, that the stars are not symmetrically arranged around the sun and that the sun moves. He began a major catalogue of the nebulae, and his son, John Herschel, continued his work. *(A.I.P./Niels Bohr Library, E. Scott Barr Collection)*

George Ellery Hale, the American astronomer who built the first large reflecting telescopes that could reveal the true nature of some nebulae to be gigantic star systems—the galaxies. Here he rests on Mount Wilson, the site of the 100-inch Hooker telescope. *(Hale Observatories)*

"The Horse's Laugh"—Harlow Shapley in a jovial mood. Shapley, through his careful measurements of the distribution of globular clusters, established the shape of our Milky Way galaxy to be a flattened disk and located its center. But he also maintained the incorrect view that all other nebulae were part of our galaxy and not other, distant "island universes" similar to ours. Evidently, it was Shapley who gave his friend the poet Robert Frost the idea for the poem "Fire and Ice." *(John Hubley)*

Edwin Hubble devoted his life to the observation of galaxies, the most distant objects known to astronomers at that time. Through his observations and those of many other astronomers, scientists became convinced that the galaxies were external star systems and that earlier theoretical speculations that the universe was expanding were correct. Under his leadership, observational cosmology became a science. Before becoming an astronomer, Hubble was a soldier, a basketball coach and a law student. (Hale Observatories, Courtesy A.I.P/Niels Bohr Library)

Subrahmanyan Chandrasekhar, in 1930, applied the concepts of the new quantum and relativity theory to elucidate the superdense matter of white dwarf stars. At the time he devised his theory, only three such peculiar stars were known to exist. Today, hundreds are known. White dwarfs are one of three possible end products of stellar evolution, the other two being neutron stars and black holes. A.I.P./ Meggers Gallery of Nobel Laureates)

Fritz Zwicky had a long, active career in astronomy. Along with Walter Baade, he conjectured the existence of neutron stars in 1933. Zwicky, through his studies of the motion of galaxies in clusters, concluded that some form of dark matter had to be present in the clusters—the first indication of the "missing mass" puzzle of the universe.

Walter Baade, who collaborated with Zwicky on the theory of neutron stars, did major observational work in astronomy. During World War II, he discovered that stars fall into two major populations—old red stars and young blue ones. Subsequently, he showed that the failure to distinguish the two star populations had led previous astronomers to underestimate both the size and the age of the universe. This reestimated, greater age for the universe, which now exceeded the age of the oldest stars, resolved one of the main objections to the big-bang theory of the origin of the universe. (Dorothy Davis Locanthy)

George Gamow learned quantum theory at the Niels Bohr Institute in Copenhagen in the late 1920s and went on to explicate "quantum tunneling"—how nuclear particles could tunnel right through nuclear barriers. An understanding of this process was crucial to an understanding of nuclear burning in stars' cores. In the late 1940s, Gamow and his two students Ralph Alpher and Robert Herman devised the modern version of the big-bang theory. Gamow loved jokes and tricks. His biggest joke was that the big-bang theory turned out to be correct. (A.I.P./Niels Bohr Library)

Yakob B. Zel'dovich and his collaborators in the Soviet Union pioneered the theoretical study of the early universe. They were the first to emphasize that the high energies and temperatures of the big bang provided a new proving ground for testing high-energy quantum-field theory. Zel'dovich is seen lecturing at a conference in Czechoslovakia. (Leo Goldberg)

Steven Hawking's mathematical investigations have had a profound impact on our current understanding of the universe. With Rodger Penrose, another mathematical physicist, Hawking proved the "singularity theorem," which implies that our universe must evolve from or into a very dense state. He pioneered the modern theory of black holes and showed how black holes are not strictly black but can radiate quantum particles—the Hawking radiation. Recently, he has been working on the origin of galaxies and the puzzle of the ultimate origin of the universe out of nothing. (Franklyn Institute)

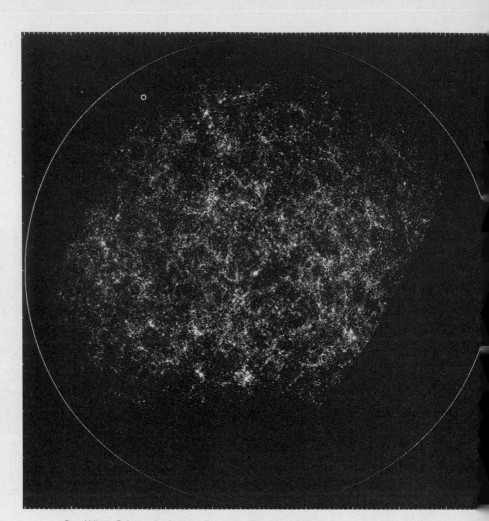

One Million Galaxies. In this two-dimensional processed illustration, representing the positions of a million galaxies in the sky, each galaxy is represented by a white dot. The line across the field is the horizon limit of the Lick telescope, where the data were taken. The white regions are clusters and superclusters of galaxies, while the darker regions are voids. Some astronomers claim there is evidence for a filamentary structure in the distribution. Recently, astronomers have made three-dimensional maps of the distribution of galaxies. The original source for this illustration is M. Seldner, B. L. Siebers, E. J. Groth and P. J. E. Peebles, *Astronomical Journal*, 82, 249, 1977.

OPPOSITE:
A variety of normal galaxies. Type SO is the lenticular galaxy, while the others are different spirals, classified according to how tightly their arms are wound up. *(Hale Observatories)*

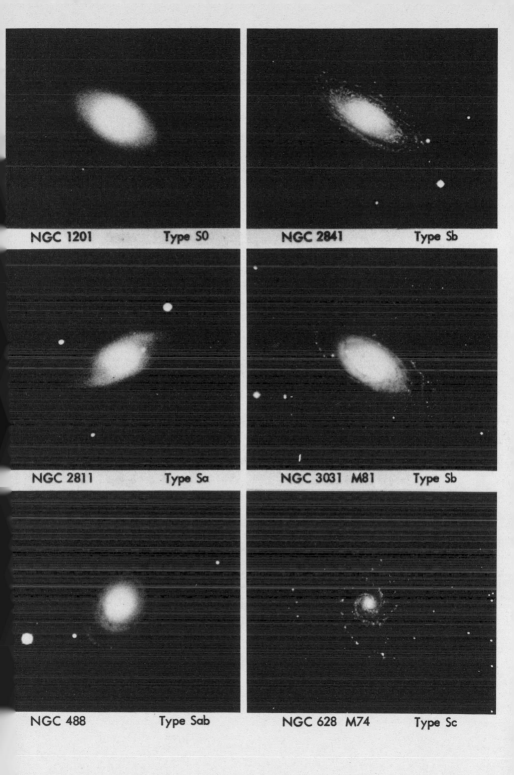

NGC 1201   Type S0

NGC 2841   Type Sb

NGC 2811   Type Sa

NGC 3031 M81   Type Sb

NGC 488   Type Sab

NGC 628 M74   Type Sc

A part of the cluster of galaxies in Coma Berenices. The Coma supercluster is one of the most dense, containing 1,300 major galaxies. *(Hale Observatories)*

The Globular Cluster, M 92. Hundreds of such globular star clusters, containing from 50,000 to a million stars, surround the disk of our galaxy, concentrating near its center. They contain old stars and are probably a clue to the formation process of the galaxy itself. *(Lick Observatory)*

Two exposures of the galaxy NGC 7331, before and during the maximum intensity of the supernova of 1959. Hundreds of such supernovas—each the explosion of a star to form a neutron star—have been observed in other galaxies, but none, in modern times, within our own galaxy. *(Lick Observatory)*

A Planetary Nebula in Ursa Major, M 97. A cloud of gas expelled into space by a dying star that was once a red giant and has now become a white dwarf. *(Hale Observatories)*

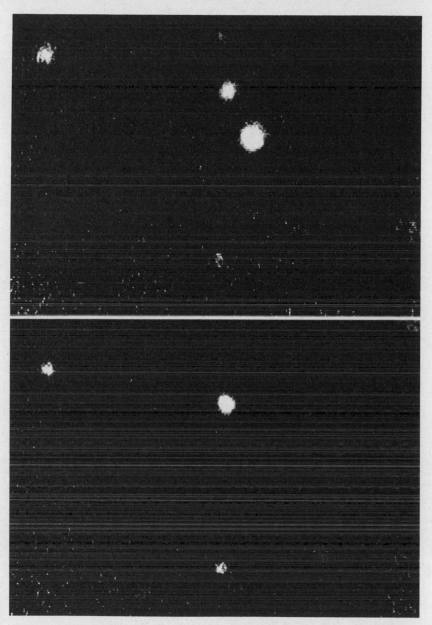

Now you see it; now you don't. Television photographs of the Crab Nebula pulsar (Baade's Star) taken at minimum and at maximum light. This pulsation of the neutron star in the visible light range was discovered by Don Taylor, John Cocke and Michael Disney. *(Lick Observatory)*

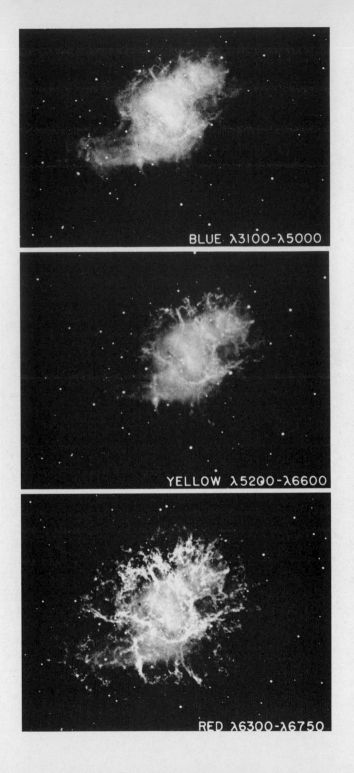

INFRARED λ7200–λ8400

OPPOSITE AND ABOVE:
The Crab Nebula in Taurus, M 1. Four photographs in blue, yellow, red and infrared light. The Crab Nebula is the supernova remnant of the pulsar (Baade's Star) near its center. *(Hale Observatories)*

The giant elliptical galaxy M 87 at the center of the Virgo cluster of galaxies. This long exposure reveals the globular clusters in the outer part. A gigantic black hole may lie at the center of this galaxy. *(Lick Observatory)*

The Horsehead Nebula (IC 434) and the bright nebulosity NGC 2024 in red light. Such huge clouds of gas and dust, located in the arms of our galaxy, are the birthplaces of stars. *(Lick Observatory)*

Two

The Early Universe

It is a task for future investigators to obtain all the properties of the universe from the laws of fundamental quantum-field physics.
—A. D. Dolgov and
Yakob B. Zel'dovich, 1981

1

The Early Universe

Thus it seems that we must reject the idea of a permanent unchangeable universe and must assume that the basic features which characterize the universe as we know it today are the direct result of some evolutionary development which must have begun a few billion years ago. . . . *With such an assumption, the problem of scientific cosmogony can be formulated as an attempt to reconstruct the evolutionary processes which led from the simplicity of the early days of creation to the present immense complexity of the universe around us.*
—George Gamow, 1951

Vannevar Bush, the distinguished statesman of science, in his essay "The Builders" compared scientific discovery and research to mining a quarry and then fitting the stones into an edifice. The stones used are varied and the whole effort seems highly unorganized, with no architect or quarrymaster overseeing the design. There is no master plan to scientific progress. Bush writes:

In these circumstances it is not at all strange that the workers sometimes proceed in erratic ways. There are those who are quite content, given a few tools, to dig away, unearthing odd blocks, piling them up in the view of fellow workers and apparently not caring whether they fit anywhere or not. . . . Some groups do not dig at all, but spend all their time arguing as to the exact arrangement of a cornice or an abutment. Some spend all their days trying to pull down a block or two that a rival has put in place. Some, indeed, neither dig nor argue, but go along with the crowd, scratch here and there, and enjoy the scenery. Some sit by and give advice, and some just sit.

The sense one gets from this metaphor is that in spite of the haphazard procedure, the edifice of science is getting built and may someday stand as a finished structure.

But if we instead compare the development of science to the evolution of life on earth, then we are comparing it to a process that is never finished. Scientific research resembles evolution in its responsiveness to altering environmental conditions, opportunism, attention to details and especially its peculiar blindness as to where it is headed. In the view of some people, human culture, of which science is a part, is but a continuation of the evolution of life to the realm of symbols and ideas, and ideas, like species, seek survival in altering environments.

Examining the evolution of the natural sciences, we are struck by the symbiosis of astronomy and physics—each a distinct discipline, yet each enhanced by the contribution of the other. Astronomy, by far the oldest science, began millennia ago with the careful observation of the heavens, an activity that continues to this day. But today the work of observational astronomers is supplemented by that of astrophysicists, who make detailed mathematical models that endeavor to account for the observations.

Physics is itself a relatively young science concerned with finding the laws of matter, space and time irrespective of whether they apply to the motion of stars or the molecules in our bodies. The laws of physics discovered here on earth can, in a sense, leap to the heavens and be applied to the most distant galaxies and quasars; the laws discovered today apply to the distant past and future as well. Physical law is universal—it is fact, never contradicted by any observation. And because of that universality, astronomy, which studies the heavens, and physics, gleaned from our experience on earth, developed a close and intimate relationship.

The fact that the entire universe is governed by simple natural laws is remarkable, profound and on the face of it, absurd. How can the vast variety of nature, the multitude of things and processes all be subject to a few simple, universal laws? Isaac Newton discovered the answer. In his formulation of mechanics we saw for the first time a clear conceptual separation between the "initial conditions" of a physical system and "the laws of motion." If we are given the initial conditions of a physical system such as the positions and momenta of billions of particles, conditions that could be arbitrarily complicated, then the laws of motion precisely determine the subsequent development of this system in time. The world was thus divided into two components: the initial conditions, which represented the complicated state of the world, and the simple universal laws that determined its subsequent development. Seldom has an idea had such wide-ranging and profound consequences.

Newton and his successors applied his laws of motion and gravity with ever-mounting success to the motions of the moon, planets and comets. The domain of the law of universal gravitation was even extended into the distant stars through William Herschel's studies of binary star systems. Because of the success of Newtonian physics in elucidating the motion of the heavens, physicists developed confidence that mathematical methods based on universal natural laws would become the most powerful conceptual tool for investigating the cosmos. That confidence was dramatically vindicated in 1846 by the discovery of a new planet, Neptune, based on mathematical studies of perturbations in the orbit of Uranus which predicted its location.

A beautiful further confirmation of the universality of natural laws and the unity of the universe was William Huggins' development of visual stellar spectroscopy and Henry Draper's photographing the hydrogen-absorption spectrum of the star Alpha Lyrae (Vega) on August 1, 1872. Huggins, an English amateur astronomer, was inspired by the spectroscopic discoveries of Kirchhoff and Fraunhofer's first observation of a stellar absorption spectrum in 1823, and he went on to observe many stellar spectra; but this work was more of an art than a science until photography was introduced. Draper was dean of the medical school at New York University and a member of a distinguished American scientific family. He spent many years patiently designing and building a 28-inch reflecting telescope at his observatory at Hastings-on-Hudson and developing the nascent photographic techniques required to photograph the dim star spectra. No one had done it before. These major achievements by Huggins and Draper established forever that the matter of a star is composed of the same atoms that we find here on earth. Astrophysics became a mature science.

The symbiosis of astronomy and physics, while mutually enhancing during the period dominated by classical, Newtonian physics, blossomed during the twentieth century with the advent of the new quantum theory of atoms in 1927. Overturning the earlier Newtonian system, quantum theory was a new mechanics radical in its implications for reality, comprehensive in scope and strangely but sublimely coherent. Physicists moved on to successfully apply the new theory to chemistry, nuclear physics, solid-state matter and the subnuclear world of elementary particles. They also knew that new quantum concepts did not apply only to terrestrial atoms; they were universal and applied to distant stars as well.

The new quantum mechanics realized some of its greatest triumphs by elucidating the outstanding puzzles of astrophysics. Chandrasekhar, using the rules of the new quantum theory, first understood the bizarre matter of the dense companion of Sirius, launching the modern theory of white dwarfs. Hans Bethe, Carl von Weizsäcker and many others working on the theory of nuclear burning in the interior of stars pioneered modern astrophysics. Neutron stars, or pulsars, are made of matter not even seen on earth; to understand them involves studying the subnuclear particles encountered in high-energy physics laboratories.

Sometimes, as a theoretical physicist, I like to provoke my friends in astrophysics by asking Why are they having such difficulty understanding the properties of stars—how they are born, evolve, burn out and explode—when the needed basic laws of physics are fully understood? But this is a bit like asking a biologist to explain the properties of a cell starting from the rules of quantum chemistry— the complexity of the undertaking is forbidding. And such complexity is an essential complication inherent in the very nature of the object of investigation.

This complexity arises because of a kind of "causal decoupling" between different organizational levels as one moves from the microcosm to the macrocosm. For example, to understand chemistry one must comprehend the rules obeyed by the valence electrons in the outer parts of atoms. The details of the atomic nucleus—the quarks inside the protons and neutrons—are "causally decoupled" from the chemical properties of the atom. Another example from molecular biology of this "causal decoupling" is the fact that the biological function of proteins is decoupled from how they are coded for in the genetic material. Science abounds in examples of this "causal decoupling"—an important separation between the material levels of nature which becomes reflected in the establishment of separate scientific disciplines.

We thus see that it is one thing to know the basic microscopic physical laws, and quite another to intuit the implications of those laws for the macroscopic system one wants to understand. While knowledge of the basic laws of physics can help astrophysicists, their ability to ask the right questions—for instance, what processes are important, which can be ignored, and what are the crucial features of this particular system—is a creative skill in its own right.

No one doubts today that the laws of physics, gleaned from terrestrial experiments, will provide the foundation of a complete theory of stars. Yet even in the face of such success it is clear that the traditional relation between physics and astronomy cannot continue as it has been in the past. The relation must, in fact, become more intimate. The reason is that observational astronomers are now exploring the quasars, the cores of galaxies and the big bang. Each is characterized by processes of such intensity that terrestrial physics experiments cannot match them. Hence to test their high-energy theories, physicists can be guided only by astronomical observation. The whole universe now becomes the unique proving ground for the laws of physics. Nothing less will do, because it is the universe in its most extreme conditions that physicists endeavor to understand.

The necessity for a more intimate relation between astronomy and physics becomes dramatically apparent if we examine the big-bang theory of the origin of the universe. According to this theory, if we were to go backward in time the temperature of the universe would increase, possibly without limit. Temperature is a measure of the energy of motion of particles—in this case quantum particles. What, then, are the physical laws that govern the interaction of quantum particles at these ultrahigh energies?

No one can answer this question with certainty. It would be exciting if physicists could check their theories by creating for a fraction of a second in an accelerator laboratory the conditions that prevailed in the very early big bang. The largest existing accelerators effectively take us back to times when the universe was only one-billionth of a second old—an immense accomplishment. But the answers to important questions about the universe that physicists are now asking depend on its properties *before* it was one-billionth of a second old. Such early times are beyond the reach of any practical accelerator experiments. Hence in order to explore the very early universe and answer questions about its very origin, physicists and others are required to adopt a different attitude. Rather than looking to the high-energy accelerators for clues, they look to the "big accelerator in the sky"—the big bang and its aftermath. The whole universe becomes an "experiment" for deducing the ultimate laws of matter. This new investigative science integrates the science of the smallest things we know—the quanta—with that of the largest—the cosmos.

How do cosmologists and quantum physicists investigating the early universe proceed? They start by using the relativistic quantum-field theories of the quantum particles that explain high-energy experiments done in accelerator laboratories. Casting caution aside, they extrapolate those theoretical models to the ultrahigh energy of the very early universe. They then use these models to deduce some remarkable features of the universe—such as why it consists of matter and not equal amounts of matter and antimatter, or the existence of tiny quantum fluctuations in the primordial fireball that eventually formed galaxies and superclusters.

In some ways this enterprise of making mathematical models of the very early universe resembles the earlier enterprise of making models of the interiors of stars on the basis of nuclear physics. No one can go to the interior of a star to check these models directly, nor can one go back to the big-bang fireball to check the models of high-energy quantum particles. But with that analogy the resemblance of the two enterprises ceases; the contrasts are significant. To begin with, there are lots of stars each with different properties and in different stages of evolution, providing astrophysicists with a welter of data which constrains the mathematical models enormously. In contrast to the multitude of stars, there is only one early universe, and it cannot be observed directly. This early era left scattered fossils, contemporary clues about its properties—the galaxies, their distribution in space, the microwave background radiation, the relative abundance of the chemical elements. But the deeper difference between the modeling of stars and modeling the very early universe is that physicists have experimentally explored the laws of nuclear physics applicable to the interiors of stars, while no terrestrial experimentation seems possible for checking the laws that apply to the very early universe.

The task of cosmologists studying the early universe seems impossibly difficult. It is as if astrophysicists, instead of using the known laws of nuclear physics, were asked to deduce them from the observed properties of stars! But there is hope, and it lies in the fact that the early universe may be a simpler object than the interior of a star. Physicists believe that at those very early, hot times the interactions of the quantum particles were much more symmetrical. They expect the complexities to disappear and the physical situation to become manageable. If that is so, then cosmologists stand a good chance of understanding the very early universe. Whether this is more than a vain wish remains an open question, but current theoretical prejudice favors it.

It is worth reminding ourselves that if anyone had suggested forty years ago that physicists would know the state of the universe when it was three minutes old, he would have been laughed at. But observational data, such as the relative abundance of the chemical elements and the temperature of the microwave background radiation, provided powerful new data whose existence could not have been anticipated then. Likewise, new cosmological data may show up in the next decade that will help us solve the puzzle of the origin of the universe.

How did scientists get to this point in their research? What led them to thinking about these problems in this new way?

The first modern suggestion about the origin of the universe harks back to Georges Lemaître's postulation of a "primeval atom" at the beginning of the universe. His "atom" was similar to the big-bang fireball. In his view, the "atom" exploded into fragments and these into still smaller fragments until the universe as we see it today emerged. In 1951, he wrote, "The evolution of the world can be compared to a display of fireworks that has just ended: some few red wisps, ashes, and smoke. Standing on a cooled cinder, we see the slow fading of suns, and we try to recall the vanished brilliance of the origin of the worlds."

Although Lemaître, "the father of the big-bang theory," took the first step, the modern version of the big bang is due to George Gamow and his two students Ralph Alpher and Robert Herman. In 1948 they did calculations of the synthesis of the chemical elements in the primordial big-bang explosion and in so doing brought the big-bang idea out of the realm of speculation and into the realm of observational science. Alpher and Herman estimated the temperature of the universe today to be 5 Kelvin above absolute zero—an estimate essentially confirmed when this temperature was measured to be 2.7K some eighteen years later by Penzias and Wilson. This direct observation of the background-radiation temperature played the major role in convincing most scientists of the correctness of the big-bang idea.

Gamow's program of calculating the relative abundance of chemical elements created in the big bang lay dormant for sixteen years. Then it started up again in 1964 with the work of Peebles at Princeton, Zel'dovich in the Soviet Union—both unaware of the earlier work of Gamow, Alpher and Herman—and Hoyle and R. J. Tayler in England. In 1967, William Fowler, Fred Hoyle, Robert Wagoner and subsequent workers did refined computer calculations to show that the right amount of helium and deuterium—elements that, in the quantities observed in the universe, cannot have been made inside stars—could get made in the big bang. On the basis of these calculations, whose success depends on detailed properties of the universe when it is only seconds and minutes old, most physicists feel confident that the state of the universe at that time is rather well understood.

Some physicists wanted to push to earlier times, beyond a few seconds deeper into the big bang. John A. Wheeler, then at Princeton University, wandering the country like an itinerant lecturer, continued to remind his colleagues that the early universe "confronts physics with its greatest crisis." In the Soviet Union, Zel'dovich and his coworkers heroically pioneered this whole area, emphasizing that the final proving ground for quantum-particle theory is the very early universe. They did many calculations to support their viewpoint. In the United States, relativists, astronomers and physicists, recognizing the growing importance of relativity theory, organized the first "Texas Conference on Relativistic Astrophysics" in 1963, a series of conferences that continues to this day to include research on the early universe.

But while these developments were occurring, most quantum-particle physicists in the early 1970s were not very much interested in the big bang. As Steven Weinberg remarked, "It was extraordinarily difficult for physicists to take seriously *any* theory of the early universe." Their attention was focused instead on exciting new models of the elementary particles, such as the quark model of subnuclear matter and the idea that electromagnetic and weak interactions, usually thought of as distinct, were but manifestations of a single "unified field." With such exciting ideas to play with it is no wonder that the problems of the early universe did not interest the particle physicists. Ironically, the very success of these new field theories in accounting for high-energy physics experiments moved the physicists to press the comparison between their theories and observation even further. And here, perforce, they were led to examine the consequences of their formulae for the big bang.

Heinrich Hertz, the nineteenth-century German physicist, once wrote, "One cannot escape the feeling that these mathematical formulae have an independent existence and an intelligence of their own, that they are wiser than we, wiser even than their discoverers, that we get more out of them than was originally put into them." This power of the equations of physics to illuminate the unknown was dramatically exemplified by the application of the new theories of the quantum particles to the early universe. Weinberg, echoing the thoughts of Hertz, wrote, "This is often the way it is in physics— our mistake is not that we take our theories too seriously, but that we do not take them seriously enough. It is always hard to realize that these numbers and equations we play with at our desks have something to do with the real world." Weinberg, one of the first quantum-particle theorists to take the early universe seriously, persuaded many of his colleagues to do likewise with his influential book *The First Three Minutes*. As Zel'dovich and his collaborator A. D. Doglov remarked in a recent review article, "Many physicists owe their acquaintance with modern cosmology to the book *The First Three Minutes* by S. Weinberg (1977)."

But some physicists, in the role of conservative critics, think that theorists exploring, on paper, the very early universe have gone too far. Extrapolating from theories that work in the relatively low-energy domain examined by terrestrial accelerators to such ultrahigh energies is a dubious enterprise. The critics could well be right. Yet the noteworthy feature of this recent undertaking to comprehend the origin of the universe is not whether its findings are right or wrong —that is, of course, very important—but to realize that for the first time the problem of the origin of the universe is being discussed in rational, mathematical terms. The advent of relativistic unified-quantum-field theory has introduced new concepts into the repertoire of the theorists—concepts that may be able to account for the origin of the universe. No wonder theoretical physicists want to proceed. If they accomplish their goal, and it may take some time, then it will be one of the great intellectual feats of science. Until that time comes, we can all join and watch the explorations, frustrations and successes of their efforts.

Yet no matter how far we descend into the big-bang explosion, matter is present. How, then, can we understand the very point of origin? Where does the universe's matter come from? Do the laws of physics break down, and will we have to give up and adopt a mystical attitude?

A few physicists, abandoning all caution, press on, trying to grasp the spark from whence it all came. To do this they invoke "wild ideas," new concepts, which neither contradict any experiment nor have evidence to support them. Such ideas involve spaces of five or more dimensions, "supersymmetries" and grand-unified-field theories (GUTs). Only a few physicists have any confidence that such ideas are completely right; they are certainly untested. Yet if a rational picture of the origin of the cosmos comes out of these ideas, it could conceivably bring the science of physics to an end.

It is ironic that to understand the largest thing we know—the entire universe—requires our mastering the laws governing the smallest entities—the quantum particles. Science abounds in such ironies. Like most thinking and feeling people, scientists are sensitive to the mystery of existence. Yet, ironically, as they explore the universe they increasingly come to understand it as subject to physical law like any other material entity, though, of course, it is a unique entity requiring new concepts for its elucidation. The mystery of existence lies less in the observed material universe than in our capacity to comprehend it in the first place. And in time that too may become less mysterious.

Some people object to the purely scientific, often reductionalistic approach and instead appeal to their deep-felt intuition that the whole universe is a special unity with a law unto itself. They feel the reductionalist approach denies some vital order of being. But these preconceived attitudes cause those people to close their minds and miss out on the true excitement of ongoing scientific discovery. Through these discoveries a new world view is being created which will profoundly influence all of culture.

For millennia, humans created symbolic cultural systems that appealed to our need to feel connected to the whole of existence. Yet such holistic visions of the universe, while they serve an important purpose, are from the standpoint of the natural sciences heuristically sterile and empirically empty. What is being created by science today is a new vision of the universe and humanity's place in it, a vision devoid of any disparity between holism and reductionalism, a vision for which such a distinction ceases to be of relevance. We may yet see the universe in a grain of sand.

The first part of this book, "Herschel's Garden," described the inhabitants of our universe, the stars and galaxies, a universe discovered primarily by observational astronomers. In this part of the book, "The Early Universe," we leave the secure ground of observational astronomy and enter into the world of concepts of modern physics. While these concepts are grounded in detailed experiments, the emphasis in this part is somewhat different because our topic—the early universe—lies beyond the touch of direct observation. Instead of direct observations it is physical concepts and the reality they imply that now take the lead in guiding us through the early universe.

In order to understand how physicists view the very early universe we will first make an excursion, in the next few chapters, into "relativistic quantum-field theory"—the language of modern physics— and into the world of subatomic particles. Armed with this knowledge, and also some thermodynamics, we will then be prepared to descend backward in time into the big bang and show how the properties of particles influenced that remarkable event.

2
Fields, Quanta and Symmetry

We know many laws of nature and we hope and expect to discover more. Nobody can foresee the next such law that will be discovered. Nevertheless, there is a structure in laws of nature which we call the laws of invariance. This structure is so far-reaching in some cases that laws of nature were guessed on the basis of the postulate that they fit into the invariance structure.

—Eugene P. Wigner

Werner Heisenberg, a student of the physicist Arnold Sommerfeld in Munich, Germany, got his doctoral degree in 1924. But before his degree in mathematical physics was actually awarded he had to take the required oral exam—usually regarded as a ritual formality, but in fact the last chance for a physics faculty to deny a student entry into its professional tribe. There happened at the time to be a running feud between Sommerfeld and another faculty member. And it is often the case in such disputes that one faculty member will try to embarrass the other by picking on his students. Heisenberg, in his oral exam, was asked by Sommerfeld's antagonistic colleague to calculate the resolving power of a microscope—an elementary calculation if you know a little optics. Heisenberg, genius that he was, could not do the calculation because he didn't know the physical properties involved—much to the embarrassment of Sommerfeld. Heisenberg got his degree, but was denied the full honors that normally would have been his. He was admonished to study optics, a boring subject for someone with Heisenberg's interests. But this story has a sequel.

A year later, in 1925, Heisenberg invented matrix mechanics, the first step toward the new quantum theory of atoms. Later, working with Max Born and Pascual Jordan in Göttingen, he devised a complete version of the new quantum theory, a new dynamics which could be applied to calculation of the properties of atoms, just as Newton's mechanics had been used to calculate the orbits of planets. Although quantum mechanics—as it was later called—agreed magnificently with experiment, its creators had difficulty in interpreting it as a picture of reality. The simple visual picture of material reality that one gets from the old Newtonian mechanics (planets orbiting the sun or the motion of billiard balls) has no analogue in quantum mechanics. The visual conventions of our ordinary experience are not applicable to the microworld of atoms, and one must try to understand that world in another way.

Heisenberg and Niels Bohr struggled to find a new framework for thinking about the quantum world that would accord with the new quantum mechanics. Through his attempt to solve these interpretive problems, Heisenberg discovered the "uncertainty principle," a principle that revealed a profound feature of quantum mechanics not found in Newtonian mechanics.

According to the uncertainty principle, certain pairs of physical variables, like the position and momentum (the mass times velocity) of a particle, cannot be measured simultaneously with arbitrary precision. For example, if one repeats the measurement of the position and momentum of a single quantum particle—an electron, say—one finds that the measurements fluctuate about average values. These fluctuations are then a measure of our uncertainty in determining the position or momentum. The uncertainty principle asserts that the product of these uncertainties in the measurements cannot be reduced to zero. If the electron obeyed the laws of Newtonian mechanics, then the uncertainties could be reduced to zero and the electron's position and momentum be determined precisely. But unlike Newtonian mechanics, quantum mechanics allows us to know only a *probability* distribution of these measurements—it is inherently statistical. The way that Heisenberg illustrated this remarkable uncertainty principle was by considering the resolving power of a microscope—the very problem he had botched on his oral exam.

Suppose you look at a tiny particle under a microscope. Light strikes the particle and is scattered into the optical system of the microscope. For a specified optical system the resolving power of the microscope—the smallest distances it can distinguish—is limited by the wavelength of the light used. Clearly, one cannot see a particle and determine its position to a distance that is smaller than this wavelength; the longer-wavelength light just bends around the particle and is not significantly scattered. Hence, to establish the position of the particle to very high precision we must use light of an extremely short wavelength—at least, shorter than the size of the particle.

But, as Heisenberg realized, light can also be thought of as a stream of particles—quanta of light called photons—and the momentum of a photon is inversely proportional to its wavelength. Thus the shorter the wavelength of light, the greater the momentum of its corresponding photons. If a short-wavelength, high-momentum photon hits the particle under the microscope, it imparts some of its momentum to the particle; this causes it to move and thus creates an uncertainty in our knowledge of its momentum. The shorter we make the wavelength of the light, the better we know the position of the particle, but then the less certain we are about its final momentum. Conversely, if we sacrifice our knowledge of the particle's position and use longer-wavelength light, then we can establish its momentum with greater certainty. But if quantum mechanics is correct, we cannot determine with absolute precision both the particle's position and its momentum.

The "Heisenberg microscope," as this device was later called, illustrated the physical basis of the uncertainty principle. Heisenberg probably would have discovered his uncertainty principle even if he had never been required to study optics. But without the embarrassment at his oral exam, it is unlikely that he would have recalled such a simple, physically intuitive illustration of his mathematical ideas. The invention of the Heisenberg microscope illustrates the creative power of genius to turn its defeats into victories of another kind.

Heisenberg's microscope utilizes a feature of the quantum world that is quite general: in order to "see" the atomic quantum world we must scatter other quantum particles from the objects we want to observe. Not surprisingly, in order to explore the microcosmos of quantum particles we need small probes, and the smallest are the quantum particles themselves. Physicists explored the microworld by observing quantum-particle collisions. The higher the energy and momentum of the colliding particles, the shorter the wavelength and the smaller the distances that they can resolve. For this reason, physicists attempting to explore ever-smaller distances require machines that accelerate quantum particles to ever-higher energies and then collide them with other target particles. Some prototypic particle accelerators were built before World War II, but not until after the war did high-energy particle physics enter its heroic age. Immense high-energy accelerators were built in the United States, Western Europe and the Soviet Union for the express purpose of exploring the microworld of quantum particles. Armed with these instruments, physicists from nations that had shortly before been at war now joined forces for a common assault on the microworld, a world that none of them had ever seen before.

These machines opened a window on the world beyond the atomic nucleus—the tiny core of an atom only one ten-thousandth the size of the whole atom. The nucleus is composed primarily of two types of particles, the proton, possessing a unit of electrical charge, and the neutron, similar to the proton in many ways but with no electric charge. Protons and neutrons have very strong interactions that bind them tightly together to form the nucleus. Physicists were eager to study that strong force, for they thought the clue to the ultimate structure of matter lay therein. But no one could have anticipated the rich and complex world of particles to which this strong nuclear force gave rise, nor how long it would be before a truly fundamental theory explaining that force would be discovered. Decades of frustration lay ahead. Yet it was in the smithy of frustration and ignorance that physicists forged their confidence in the correct theory when it finally arrived.

At the beginning of these explorations in the late 1940s, physicists discovered a few more strongly interacting particles besides protons and neutrons which they called pi mesons. Then in the 1950s, as they built accelerators of still-higher energy, they found more and more of these strongly interacting particles, among them hyperons, K mesons, rho mesons, strange particles—a whole zoo of particles, probably infinite in number. All these strongly interacting particles were given the collective name of "hadrons," meaning strong, heavy, thick. Most of them are highly unstable and decay rapidly into other, more stable hadrons. What could nature possibly be telling us? This proliferation of different kinds of subatomic particles seemed like a joke. Nature, according to some unwritten belief of physicists, was supposed to get simpler at the most fundamental level, not more complicated.

Today that belief in the simplicity of nature is vindicated. The hadrons, including the familiar proton and neutron, turned out not to be fundamental, irreducible units of matter but were built up out of still smaller units—which do appear to be irreducible—called quarks. This quark model of hadron structure, proposed in 1963 by Murray Gell-Mann and independently by George Zweig, was beautifully confirmed in a series of experiments done at the Stanford Linear Accelerator Center in 1968. These experiments detected point-like quarks sitting inside the proton and neutron "like raisins in a pudding."

Physicists now view the hadrons as manifestations of the dynamics of a few quarks orbiting about each other, bound together in a tiny region of space—an immense simplification if compared with the infinite zoo of particles. In many ways this simplification was similar to that achieved by nineteenth-century chemists when they realized that thousands of molecular compounds could be built out of only about eight dozen atomic elements.

By the late 1970s, after major theoretical and experimental discoveries, a new picture of the subatomic microworld was intact. The basic units of matter were the quarks and other particles called leptons and gluons. Through the interactions of these quantum particles every material thing in the universe could, in principle, be accounted for. This was a major accomplishment in the attempt to comprehend nature. It provided the conceptual tool needed to understand the big bang.

The mathematical model that describes these particles and their interactions is called the "standard model." I will describe it in detail in our next chapter. But before I do so, it is important to find a way of envisioning the microworld of quantum particles. What kind of "stuff" are these particles made of? How can we think about the quantum world down at subnuclear distances? In order to deal with such questions physicists have invented a highly mathematical language called "relativistic quantum-field theory." It provides the conceptual framework for thinking about the interactions of quantum particles, just as Newtonian physics provides the conceptual framework for thinking about the motion of the planets.

Theoretical physicists invented relativistic quantum-field theory in the 1920s. It grew out of their attempt to make the new quantum theory consistent with Einstein's theory of special relativity. Achieving that consistency proved far more difficult than anyone had anticipated. As Steven Weinberg remarked:

> Quantum mechanics without relativity would allow us to conceive of a great many possible physical systems. . . . However, when you put quantum mechanics together with relativity, you find that it is nearly impossible to conceive of any possible physical system at all. Nature somehow manages to be both relativistic and quantum-mechanical; but these two requirements restrict it so much that it has only a limited choice of how to be—hopefully a very limited choice.

As these remarks emphasize, both the principle of relativity and the principles of quantum theory were very restrictive requirements, and it was not at all clear that they could be successfully joined together in a mathematical description of the world. But joined they were. The consequences were profound.

One of the first steps was taken in 1926 by the theoretical physicists Max Born, Werner Heisenberg and Pascual Jordan, who showed how to apply the new quantum concepts to the electromagnetic field, a field that already obeyed the requirements of Einstein's special relativity theory. Their work demonstrated how Einstein's earlier idea of the photon as a particle of light could be mathematically described.

The next major steps were taken in 1928 by Jordan and Eugene Wigner and in 1929–1930 by Heisenberg and Wolfgang Pauli. They showed that each different field—the electromagnetic field, the electron field and so on—had an associated particle. Particles were manifestations of a "quantized" field. This was the basic idea of modern field theory, which banished forever the old idea that particles and fields were separate entities. Fields were the fundamental entities, but they were manifested in the world as particles.

Theoretical physicists struggled for decades to deepen their understanding of these new field theories, an intellectual adventure that continues to this day. They were brought to these new concepts by the discoveries of their experimental colleagues—discoveries which demanded an explanation—as well as their own urge to find a coherent language to describe the quantum world. How do these new ideas lead us to think of the quantum particles?

It is difficult to keep from imagining quantum particles as ordinary things that happen to be very small. Physicists, thinking about the particles, slip into that way of thinking all the time because it is so easy. But fundamental particles are not made of "stuff" the way a chair is made of wood, screws and glue. Any such simple visual picture breaks down completely once one begins to ask detailed questions. That is when the weird world of quantum reality comes into play.

The first way that physicists think of these particles is in terms of the intrinsic properties that classify them—like their mass, spin, electric charge and so forth. The second way they think of them is in terms of their interaction with other particles. Once a physicist knows the intrinsic properties of a quantum particle and knows all its interactions, he then knows everything he can know about that particle. But how do physicists describe what they know?

The observed properties of the quantum particles can be precisely described in the language of mathematics, and within that language the notion of symmetry has come to play an increasingly important role. Why symmetry? One reason is that fundamental quantum particles like electrons or photons are believed to have no structure— they are not made out of simpler parts—but nonetheless they possess certain symmetries, the way a crystal has symmetry. Furthermore, the electron, whatever it is, is very small, perhaps a point particle. For description of something that has no parts and is very small, concepts of symmetry turn out to be extremely useful. For example, imagine a sphere sitting in space. The sphere appears the same if we move around it—it has the property of being spherically symmetrical. If the sphere is shrunk to a very small size, even to a point, this property of spherical symmetry is retained; the particle is also spherically symmetrical. If instead of a sphere we image an ellipsoid, which has symmetry about only one axis, then as this is shrunk to zero size its symmetry also is retained. We learn from these examples that even if something is very small and without structure, it still can have specific symmetries. Even to a complicated composite object such as an atom built up out of electrons and nuclei, or an atomic nucleus built up out of protons and neutrons, the concept of symmetry can be applied with great effect. The interactions between the constituents of atoms and nuclei also possess specific symmetries which help to determine the composite structure, just as the symmetries of ceramic tiles determine the patterns in which they can be laid down.

As we will see below, elementary quantum particles are defined in terms of *how they transform* under mathematical "symmetry operations": for example, how a quantum particle changes if we rotate it about an axis in space—a symmetry operation. The role of symmetry in describing the properties of quantum particles is central to the entire enterprise of modern physics. As C. N. Yang, the theoretical physicist, expressed it,

Nature seems to take advantage of the simple mathematical representations of the symmetry laws. When one pauses to consider the elegance and the beautiful perfection of the mathematical reasoning involved and contrast it with the complex and far-reaching physical consequences, a deep sense of respect for the power of the symmetry laws never fails to develop.

In order to get a better grasp on the relation between abstract mathematical symmetries and how they are represented by actual elementary particles, let us recall the nineteenth-century application of symmetry ideas to the various kinds of crystals that form in nature, like salt or diamonds or rubies. A crystal can be thought of as a spatial lattice, an actual periodic structure of atoms in space.

For the moment forget about crystals and instead imagine a mathematical lattice of points joined by lines, such as the mesh on a wire screen, filling all of space. A screen mesh is a square grid, but one could well imagine a grid made of diamond shapes or triangles, as long as it repeats periodically. Mathematicians determined and classified all such possible periodic lattice structures in three-dimensional space through the use of abstract ideas about symmetry. Here the symmetry is the symmetry or invariance one would observe by performing a displacement in space, like moving along the edge of a cube in an infinite cubic lattice, and finding that the lattice has not changed. Such symmetries can thus be viewed as abstract mathematical operations in three-dimensional space.

If we now return to thinking about the actual crystals we find in nature, then all possible real crystals are representations of these mathematical symmetries because they too are periodic structures in space. Likewise, the abstract mathematical symmetries of the laws of nature are represented by the actual quantum particles observed in nature—electrons, protons and neutrons. Very loosely speaking, quantum particles are like microscopic crystals and can be described completely in terms of their symmetry properties. Symmetry is the key that opens to human understanding the door to the microscopic world.

Relativistic quantum-field theory—the language describing the symmetries of quantum particles—is a complex mathematical discipline, but its basic ideas can be grasped in elementary terms. Let us begin by examining the meaning of each of the words "relativistic," "quantum" and "field"—all adjectives of the word "theory," meaning a picture of reality. The remainder of this chapter deals with these mathematical concepts, which although helpful for understanding the origin of the universe, are not completely essential for the general reader. It might be omitted on a first reading.

RELATIVISTIC

"Relativistic" refers to Einstein's 1905 theory of special relativity. Einstein found the correct laws of space and time transformation that govern distance and time measurements carried out by observers moving relative to each other. If the relative velocity of the two observers was small compared with the speed of light, then the Einstein space-time transformations corresponded to those of Newtonian mechanics. But for large relative velocities—comparable to the velocity of light—Einstein's space-time transformations were different from the Newtonian ones and implied a radically different picture of space and time. Subsequently, Einstein's transformations were mathematically understood to correspond to rotations in Minkowski's four-dimensional space-time. Any theory of the fundamental particles has to obey Einstein's space-time–transformation rules—a profound and extremely restrictive requirement on the mathematical description of the particles. These rules, on how space-time transforms from one observer to another, are associated with what mathematicians call a "symmetry group." But what is a symmetry group?

Symmetries are seen all around us—the approximate bilateral symmetry of our bodies, the spherical symmetry of a ball, the cylindrical symmetry of a food can. A symmetry has to do with how some objects remain unchanged or invariant if we transform them. For example, if we rotate a perfect sphere about any axis or a cylinder about its axis it remains unchanged—a manifestation of its specific symmetry. Such operations are called *symmetry operations*. Early on, pure mathematicians such as Hermann Weyl, who also learned the new quantum theory, emphasized that symmetries could have a profound influence on the elucidation of quantum problems.

By the nineteenth century, mathematicians had already invented the mathematical description of all such possible symmetry operations in terms of a new discipline called group theory. A basic idea of group theory is to symbolically describe symmetry operations such as rotations by using algebra. For example, suppose we denote the rotation of an object about some specific axis labeled 1 through a specific angle by R_1. Let R_2 and R_3 denote other rotations through other angles about different axes labeled 2 and 3. Then, if we algebraically write the product $R_2 \times R_1$, then this means: first perform the

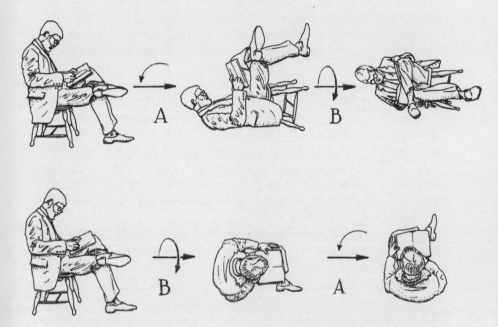

Rotational operations do not obey the algebraic commutative rule A × B = B × A. In this example, A corresponds to a counterclockwise rotation of 90 degrees about an axis perpendicular to the plane of the page and B corresponds to a rotation of 90 degrees about a horizontal axis. By rotating the reader with the B operation and then the A operation, we see that the final result is not the same as that obtained when the operations are done in the opposite order.

rotation R_1 and then follow this operation by the rotation R_2. The joint operation $R_2 \times R_1$ is itself a rotation. Notice that for rotations, $R_2 \times R_1$ is not equal to $R_1 \times R_2$; but we do not assume that here.

Next suppose we perform the rotation R_3, so the resulting rotation is now $R_3 \times (R_2 \times R_1)$, which means the rotation $R_2 \times R_1$ followed by R_3. Suppose we now begin again and perform R_1 and follow this by the joint rotation denoted by $R_3 \times R_2$, so the net result is $(R_3 \times R_2) \times R_1 = R_3 \times (R_2 \times R_1)$; we find that the "associative law" holds for these rotational operations. This rule is one of the axioms of group theory.

Furthermore, we notice that there is a rather simple rotation of the object which corresponds to leaving it unchanged—the identity operation denoted I, which means perform no rotation at all. Clearly $I \times R_1 = R_1 \times I = R_1$. The existence of the identity operation I is the second axiom of group theory.

Finally, we assume that there is an inverse operation which can undo any rotation, corresponding to just rotating the object backward. The inverse operation to the rotation R_1 is denoted by R_1^{-1}, and it has the property $R_1 \times R_1^{-1} = I = R_1^{-1} \times R_1$.

From these three deceptively simple axioms—the associative law, the existence of the identity and an inverse—emerges the beautiful structure of mathematical group theory, much in the way that the beauties of plane geometry emerge from Euclid's axioms. Although we have illustrated the algebraic axioms of group theory in terms of rotations of an object in three-dimensional space, the axioms are far more general and apply to many kinds of symmetry transformations in multidimensional spaces—the interchange of objects, spatial reflections and so on. Formidable algebraic methods can be brought to bear on the notion of symmetry, and all such possible symmetries were classified and studied by mathematicians. But what do such abstract mathematical ideas have to do with physics?

Imagine two physicists at different locations in space, both of them observing the same object in yet a third location. The two physicists make measurements on this object relative to their own positions and

---

OPPOSITE:
Two different physicists, at relative rest in space, holding on to their respective coordinate systems, represented by three mutually perpendicular arrows. If they want to communicate the results of measurements carried out relative to their coordinate systems, they have to know how the two systems are related.

The most general coordinate transformation that will transform one system into the other is a translation of the origin point of one coordinate system to the other, followed by a rotation.

then decide to communicate their results to each other. Since each physicist made the measurements relative to his own system of measurement coordinates, in order for them to communicate they need to transform or translate the measurements made in one coordinate system to those made in another. The most general such transformation of the coordinates for two physicists at rest relative to each other (as we assume here) is a translation (a straight-line displacement in space) and a rotation about an axis. It is easy to convince oneself that any such translations and rotations when described algebraically obey the axioms of group theory. Group theory and symmetry, we see, come into consideration the moment we ask how various measurements made in different coordinate systems transform one into another—the general laws of space and time transformations.

It is important to realize that symmetry concepts apply to the general *laws* of physics, not to specific *events* or configurations. In the example I gave, it is important that *any* two systems of coordinates (not just some) can be transformed one into the other by a translation and a rotation. Furthermore, if any two physicists using different coordinate systems deduce the same laws of physics, then we may conclude that the laws of physics are translationally and rotationally invariant—they apply indifferently to the place that one is located or one's orientation in space. Symmetries of the laws of physics thus express invariances.

So far we have considered translations and rotations in ordinary three-dimensional space. But with a bit of reflection one can grasp that these same ideas must generalize and apply to the four-dimensional space-time of Minkowski, which is relevant for Einstein's laws of space-time transformation between moving observers. Physicists have learned that the deeper content of Einstein's special relativity theory is that the laws of physics are invariant only for symmetry operations corresponding to rotations and translations in four-dimensional space-time. If we impose this symmetry requirement (which is the same as requiring that special relativity be valid), then we discover something quite remarkable.

Eugene Wigner, the Princeton physicist, was among the first to explore this "symmetry group" aspect of the Einstein transformations and apply it to quantum particles. In 1939 he wrote a paper that showed how these purely mathematical considerations of group theory imply that quantum particles can be classified—a remarkable and profound development. In some ways Wigner's accomplishment resembles that of the earlier generation of scientists who classified all the possible crystals through the use of symmetry groups, the so-called "crystal groups," of periodic spatial lattices. While crystals can be represented on spatial lattices, objects such as quantum particles (or for that matter, any object, since it exists in four-dimensional space-time) must be representations of the corresponding symmetries of space-time embodied in the Einstein transformations. Wigner shows that this leads to the classification of quantum particles.

First Wigner showed that every quantum particle could be classified according to its rest mass. If the particle was moving and the particle's rest mass was not zero, then you could imagine catching up to the particle, so that relative to your own motion it would now be at rest, and measuring its rest mass precisely. On the other hand, if the rest mass of a particle was exactly zero—like the photon's, the particle of light—it would always move at the speed of light and you could never catch up to it. So all particles could be classified according to their rest mass, whether it was zero or not.

Wigner's work also allows for "tachyons"—hypothetical particles that *always* move faster than light. Tachyons have never been observed, and no one has ever succeeded in formulating a consistent mathematical theory of interacting tachyons. Gerald Feinberg, the physicist who named the tachyons, once commented to me that the only place "tachyons" could be found is in Webster's dictionary.

Wigner's second important principle of classification is that every quantum particle must have a definite spin. One might imagine the particles as little spinning tops. This spin, in special units, could have only the values 0, ½, 1, ³⁄₂, 2, ⁵⁄₂, 3 . . . —either an integer or a half-integer value; it was quantized. If a particle were ever discovered with a spin of ⅙, it would imply a violation of special relativity and be a serious overthrow of the laws of physics.

Particles of integer spin, 0, 1, 2 . . . , are called "bosons," while particles of half-integer spin, ½, ³⁄₂, ⁵⁄₂ . . . , are called "fermions"—a distinction of major importance because each set of spinning particles interacts with other particles very differently. For example, the total number of fermions entering a reaction has to equal the total number leaving—fermions are conserved. But no such conservation law applies to bosons.

From the viewpoint of quantum theory, the significance of Wigner's 1939 classification system lay in the fact that the various properties he used to classify the particles—their mass, spin and so forth

—were not subject to the Heisenberg uncertainty principle. One can measure the mass and the spin of a particle simultaneously with complete precision. Hence such properties (but not others) have unambiguous values for each particle; they may be thought of as the attributes of quantum particles.

Wigner based his work on the idea that the Einstein transformations were a symmetry group of Minkowski's space-time—one of the first fruitful applications of symmetry principles in modern particle physics. It was an especially useful idea when applied to multiparticle systems—for example, the atomic nucleus, composed of many protons and neutrons. The importance of Wigner's idea lay in the fact that once one imposed the algebraic requirement of the symmetry group on a mathematical description of the world, one automatically implied not only that the principles of special relativity would be obeyed but also that the particles in that world could be simply classified. From a single requirement, a rich structure of implications flowed.

I heard the following anecdote about how Wigner first came to apply group theory to the problems of atomic physics. Wigner and John von Neumann, the eminent mathematician, had gone to high school together in their native Budapest. They were roommates and close friends during their later studies in Berlin. Wigner was struggling with the problem of a quantum-mechanical treatment of many spinning particles and presented the problem to Von Neumann—a mathematical genius by most people's judgment. Von Neumann later made many fundamental contributions to mathematics and invented the basic concepts of the programmed computer. In the Pentagon (he was often consulted about defense problems) it was said that he was worth at least one American division.

Von Neumann grasped Wigner's problem immediately, stared at the wall and began mumbling to himself—he was thinking. After a long while, Von Neumann finally looked up at Wigner and asked, "Have you ever heard of Schur's lemma?" Schur's lemma is one of the fundamental results of group theory and was the clue to solving Wigner's problem in quantum mechanics. Evidently it was Von Neumann who helped steer Wigner into the mathematics of group theory.

FIELDS

Next let us turn to the "field" concept, which was developed in the nineteenth century, well before the invention of quantum mechanics or special relativity theory. The most familiar fields are physical entities like the electric or magnetic fields which can make their presence felt in everyday life. They are invisible and yet influence matter; a magnetic field attracts iron, for instance. Today physicists believe that all quantum particles—electrons or quarks—are manifestations of different kinds of fields. But what are fields?

Imagine a large volume of air, like the air mass over a continent. At every point in the volume of air we can assign a single number which corresponds to the temperature of the air at that point. Air temperature illustrates what physicists call a "scalar field"—a numerical function that expresses a magnitude (the air temperature) which varies from point to point in space. We can also suppose that this temperature field is a function of time; the temperature changes continuously hour by hour.

Other kinds of fields are also possible. For example, suppose that the air is moving, as it usually is. Then at every point in the air we can specify a vector, a mathematical object with both a magnitude, which expresses the speed of the air at that point, and a direction, which is the direction in which the air is moving at that point. One may imagine a vector as an arrow attached to each point in space. The velocity of air throughout space is an example of a "vector field" —it has both magnitude and direction, and it can also change over time.

Fields such as the temperature and velocity fields of air can be static and not move, or move slowly, or move in such a way that a wave field is propagated in the medium. Mathematically, the movement of fields in space and time is described by a set of "field equations."

Fields can also interact with each other. For example, if the temperature is low in some region, then hotter air will begin to move into it; the scalar temperature field thus influences the vector velocity field, and vice versa.

Physicists in the last century knew about fields like the scalar and vector fields I have just described for air. Each field necessarily had an associated medium, and the temperature field was the temperature of the air medium. Wave fields always propagated in a medium, the way sound waves propagate in air. Fields without a medium to support them seemed impossible.

James Clerk Maxwell, the nineteenth-century Scottish physicist who first wrote down the equations describing electric and magnetic fields and showed that light was an electromagnetic wave field, also thought about the question of what fields were. He designed mechanical models of the electromagnetic field—machines made of gears and screws that imitated the properties of the field. Maxwell was ambivalent about whether the electric and magnetic fields needed the medium of the "ether" that was believed to pervade all of space. Many physicists who believed in the ether tried to guess its properties from the properties of light as it propagated in this strange medium. But Einstein in his 1905 paper on special relativity showed that if he was right, any attempt to detect such an ether must fail—it was a superfluous concept. Electromagnetic fields required no medium in which to propagate, and in this sense they were fundamental and irreducible entities. Unlike the temperature and velocity fields for air, which could be reduced to the properties of moving atoms of air, the electromagnetic field had no "atomic" parts.

Today, as a consequence of Einstein's work, physicists' attitude toward fundamental fields has changed completely. Such fields are not to be explained in terms of something else like an ether. On the contrary, fundamental fields (of which there are many besides the electromagnetic field) are the primary entities in terms of which we attempt to explain everything else. As Steven Weinberg expressed it, *"the essential reality is a set of fields . . .* all else can be derived as a consequence of the quantum dynamics of those fields."

It is as meaningless to ask what fields are made of as it is to ask about the "stuff" out of which quantum particles are made. The current view is that fields are irreducible—they have no parts; they are the simplest things. Fields, like the electromagnetic field and the other fields we will encounter, are physical entities which are simply defined in terms of the field equations that describe how they change and are classified in terms of how they transform under different symmetry operations and how they interact with other fields. Once one has specified these properties of a field, it is precisely defined.

What kinds of fields are there? If we require that the fields obey Einstein's special relativity theory, then we can use Wigner's classification system. As I will subsequently discuss at length, every field corresponds to a distinct quantum particle with a specific mass and spin—the basis of their classification. Some fields correspond to massless quantum particles. These fields, which include the electromagnetic and gravitational fields, are long-range; they stretch out over long distances, so that we can easily detect their presence. Other fields describe the interactions of massive quantum particles. Such fields are quite short-range; they extend into space only over microscopic atomic or subnuclear distances.

By examining how fields transform if we rotate them, we can assign them a spin. Not surprisingly, the spin so assigned corresponds to the actual spin of the quantum particle associated with the field. The electromagnetic field has spin one, the same spin as the photons; the Dirac field has spin one-half, the same as the electron, and other kinds of fields have spin zero or three-halves or two. Wigner's work leads to a classification of every kind of field in terms of its mass and spin.

Fields also have other properties that help classify them. Among these properties are different kinds of charges, such as electric charge. Just as the property of spin for a field was related to its space-time symmetry, so too the charges of fields are related to additional symmetries called "internal symmetries." How can we think about these additional "internal" charge symmetries? What are they?

So far we have been discussing single fields with a specific mass and spin. But imagine several fields each with exactly the same mass and spin. In this case physicists continue to speak of a single field but a field with several "internal" components. The basic idea of an internal symmetry is that its operation transforms the several field components into one another in such a way that the physical situation remains unchanged.

In order to illustrate this, imagine two fields of the same kind permeating all of space and label one field the "red" field and the other the "blue" field. Using colors for the labels is of no significance; one could as well numerically label the fields 1 and 2. In analogy with the temperature field of air, let us suppose that at point x in space one has a "red temperature" $T_R(x)$ and a "blue temperature" $T_B(x)$, which are the magnitudes of the two fields at point x. However, we will suppose that the energy of the two fields depends only on the quantity $T(x)$, which is given by the formula $T^2(x) = T_R^2(x) + T_B^2(x)$—that is, the square of T is the sum of the squares of T_R and T_B.

One might go on to imagine that "red" and "blue" are labels of axes in a two-dimensional "internal" space (which has nothing to do

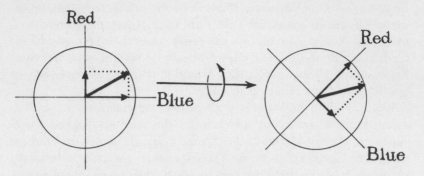

The magnitude of the "red" and "blue" fields at a point in space is indicated by the length of the arrows on the red and blue axes of an imaginary "internal" space. When the axes of this internal space are rotated, the magnitudes of the red and blue fields change. But the total field energy, which depends only on the length of the radius, does not change. This implies an "internal" symmetry of the field components.

with real physical space) and that the magnitude of the red field and the magnitude of the blue field throughout real space are measured on the corresponding "red" and "blue" axes in the internal space. A rotation of the axes in this imaginary internal space—an internal symmetry operation—changes the relative amount of the red and blue components of the field but leaves the quantity $T(x)$ unchanged because it is the radius of a circle, which does not change if the axes are rotated.

Suppose that we perform such a mathematical rotation transforming the red and blue components into each other. But such a rotation leaves $T(x)$, and hence the total energy, unchanged. The physical situation described by the field equations therefore also remains unchanged—the interactions of the two component fields are indifferent to the amount of rotation of the red and blue fields. Then we have a new symmetry—the world is unchanged by rotations in this internal space of the field components. What does this mean?

Physicists know that invariances in symmetry operations, such as the rotation we've just described, imply the existence of conserved quantities, like electric charge, which are associated with the multicomponent field. The reason for this is not difficult to grasp. A symmetry implies that something does not change—an invariance of the world. Invariance implies conservation of something, and in the case of the internal symmetries it implies the conservation of various charges. We learn that the symmetries of multicomponent fields imply that the fields possess charges that are conserved in their interactions. It was the mathematician Emmy Noether who made this relation of symmetry to conservation laws mathematically precise, thus providing one of the major reasons that theoretical physicists seek new symmetries.

Emmy Noether, whose work was so important in this regard, was the first woman recommended for a permanent faculty position at Göttingen University back in the first decades of this century. David Hilbert, one of the great mathematicians of all time, supported her. A debate about the appropriateness of Noether's appointment in the all-male faculty broke out during a faculty meeting, but Hilbert remained silent. Finally, after no resolution had been achieved, the faculty turned to Hilbert for his opinion, and he responded by reminding his colleagues that they were the faculty of a distinguished German university and not "a swimming-pool club" (which would have been sexually segregated in those days). Noether got the position and title, but not a salary. She had to leave Germany later, a refugee from the Nazis.

While multicomponent fields with "internal symmetries" may interact and get scrambled around, their associated charges never change. Hence these charges—a consequence of symmetry—imply yet another permanent label by which fields can be classified. For example, if you told a physicist that a field had a mass of 0.51 million electron volts, a spin of one-half and an electric charge of minus one, he would recognize this field as the electron field.

The internal space, instead of being just two-dimensional as it was for the "red" and "blue" fields, can have many dimensions, corresponding to many field components. Instead of a simple rotation of axis in the two-dimensional plane, the transformation can be very much more complicated, but the basic idea remains the same: if the components of a multicomponent field can be transformed into one another without alteration of the field interactions, then a symmetry is present along with an associated law of charge conservation.

We already see what an important role symmetry plays in our understanding of fields. Fields are actually defined by how they transform in various symmetry operations. Fields are not ethereal substances that pervade space and move in time; they are irreducible entities that possess specified mass, spin and charge—all properties defined by symmetry operations. Once these properties are specified, you have also completely said what a field *is*.

The classical field concept is one of the deep ideas of modern science. It provides a symbolic, mathematical language for describing the real physical world, a language which, when fully grasped, leaves no room for a further reduction in meaning. If the field concept is to be transcended, as it may be in some future time, then this will require a profound alteration of our concepts of space, time and symmetry. Today field theory is the language physicists use in talking about the fundamental material order of the cosmos.

QUANTUM

So far I have described the modern field concept purely in terms of "classical" fields—quantum concepts have played no role. But what do such fields have to do with the quantum particles—the quarks, electrons and other particles—out of which the world is actually made? Physicists discovered the answer to this question when they imposed the principles of quantum theory on the classical field concept. They learned that every field if "quantized"—made to obey the requirements of the quantum theory—describes an associated quantum particle. The quantum associated with the classical Maxwell electromagnetic field was the photon—a particle of light; the quantum associated with the classical Dirac field was the electron. In this way the distasteful dualism of particles and fields was overcome. The way in which quantum particles are classified—their mass, spin and charges—is identical to that of their associated fields.

Quantum theory also supplied an interpretation for the classical field: the *intensity* of a field at some point in space was equal to the *probability* of finding its associated quantum particle at that point. Fields were probability waves for their quantum particles. If the field was intense at some point then it was more likely that its quantum particle would be there. This "statistical interpretation" of quantum theory implies an essential indeterminacy in the laws of physics, because the *distribution* of quantum events is absolutely determined by the equations of quantum theory; *individual* events are not. For example, the theory does not specify where on a screen a particular photon passing through a hole will strike; only the distribution of many such hits can be specified precisely.

With the quantum theory successfully applied to field theory, the major puzzles that confronted physicists for the first decades of this century were solved. A powerful mathematical tool, a set of deep concepts that opened an unanticipated vista on reality fell into the hands of physicists. A new world order emerged.

The world according to this view is a vast arena of interacting fields manifested as quantum particles flying about and interacting with each other. Experience has demonstrated that this abstract mathematical description is able to correctly describe the microscopic material world as observed in the laboratory. Relativistic quantum-field theory represents the culmination of decades, if not centuries, of scientific work, and so far it has shown remarkable endurance. Its basic tenets have been challenged, but never overthrown.

The basic laws of relativistic quantum-field theory were intact by the 1930s. Since then these ideas have been enhanced, amplified and applied to the real world of quantum particles. I will mention a few of the salient developments that will guide our thinking when we turn to describing the origin of the universe.

ANTIPARTICLES

One of the first successes of relativistic quantum-field theory was the prediction of antiparticles—new quanta which were a through-the-looking-glass version of ordinary particles. Antiparticles have the same mass and spin as their partners among the ordinary particles, but their charges are reversed. The antiparticle of the electron is called the positron, and it has opposite electric charge to the electron's. If you bring electrons and positrons together they annihilate, releasing the immense energy in their mass according to the Einstein mass–energy equivalence.

How did physicists predict the existence of antiparticles? Recall that the "statistical interpretation" implied that the intensity of a field established the probability for finding its associated particles. So one can think of a field at a point in space as describing the creation or annihilation of its quantum particles with a specific probability. If this mathematical description of the creation and annihilation of quantum particles is carried out in the context of relativistic quantum-field theory, one finds that one cannot have the possibility of creating a quantum particle without also having the possibility of creating a new kind of particle—its antiparticle. The existence of antimatter is simply forced on one by the requirements of a mathematically consistent description of the creation and annihilation process in accord with both quantum and relativity theory.

The necessity for the existence of antiparticles was first grasped by the theoretical physicist Paul Dirac, who also made many other major contributions to the new quantum theory. He found the relativistic equation which now bears his name, that is obeyed by the electron's field—an accomplishment comparable to Maxwell's discovery of the electromagnetic-field equations.

When Dirac solved his equation, he found that besides describing the electron, the equation had additional solutions that described another particle with an opposite electric charge to that of the electron. What could that mean? At the time Dirac made this observation, the only known particle with this property was the proton. Dirac, not wanting to proliferate the known particles, suggested that the additional solutions to his equation described the proton. But after more careful investigation, it became clear that the particles described by the extra solutions had to have precisely the same mass as the electron; this ruled out the proton, which has a mass at least 1,800 times the electron's mass. The extra solutions therefore had to correspond to a completely new particle with the same mass as the electron but of opposite charge—an antielectron! This was beautifully confirmed experimentally when Carl Anderson, a Cal Tech physicist, actually detected the antielectron, now called the positron, in 1932.

The advent of antiparticles changed forever the way physicists thought of matter. Matter was previously thought to be permanent and immutable. Molecules could be altered, atoms could decay by radioactive processes, but the fundamental quanta were thought to be unchanging. But with Paul Dirac's discovery of antimatter, this view had to be replaced. Heisenberg put it this way:

> I believe that the discovery of particles and antiparticles by Dirac has changed our whole outlook on atomic physics. . . . Up to that time I think every physicist had thought of the elementary particles along the lines of the philosophy of Democritus, namely by considering these elementary particles as unchangeable units which are just given in nature and are always the same thing, they never change, they never can be transmuted into anything else. They are not dynamical systems, they just exist in themselves. After Dirac's discovery everything

looked different, because one could ask, why should a proton not sometimes be a proton plus a pair of electron and positron and so on? . . . Thereby the problem of dividing matter had come into a different light.

The mutability of matter became a cornerstone of the new particle physics. The fact that particles and antiparticles can together be created out of the vacuum if we supply sufficient energy is important not only for understanding how particles are created in high-energy accelerators but also for discovering the quantum processes that took place in the hot big bang.

Paul Dirac is famous for his reticence; he hardly ever speaks—but when he does, what he says always goes to the heart of the matter. I heard the following story about Dirac, which although unconfirmed is certainly believable. Richard Feynman, one of the inventors of quantum electrodynamics, who like most people likes a good conversation, was seated next to Dirac at dinner. Dirac did not talk, and finally Feynman, perhaps out of desperation, asked him, "Did you feel good when you wrote down that equation?"—referring, of course, to the eternal Dirac equation. After a long pause, Dirac said, "Yes." Then, after a still longer pause, Dirac queried almost innocently, "Are *you* working on an equation too?" Even Feynman, brilliantly inventive as he is, had no response.

RENORMALIZATION

Although the ideas of relativistic quantum-field theory successfully predicted the existence of antimatter, theoretical physicists in the 1930s and 1940s found lots of mathematical difficulties and problems with these new ideas. If they calculated quantum interaction processes using these new ideas they obtained infinite numbers, so clearly something was going wrong. Nature does not have physical quantities that are infinite. The trouble lay with the very idea of a wave field oscillating in space. No matter how small a volume of space one examines, some very short wavelengths of the field are always present, and the continued presence of those infinitely many very short waves was directly responsible for the infinite numbers the physicists were calculating. Some physicists thought that field theory might be wrong.

Yet others continued to struggle with this problem and eventually managed to tame these infinities by a mathematical trick called the "renormalization procedure." They showed that the infinite numbers appeared only in calculations of a few quantities like the mass or electric charge of the quantum particles involved, and that if these quantities were redefined, or "renormalized," by subtracting an infinitely large number, they would then yield finite predictions for all experimentally measurable quantities. Subtracting these infinities seemed like a mathematical trick; but it worked.

By the late 1940s, theoretical physicists, foremost among them Freeman Dyson, Richard Feynman, Julian Schwinger and Sin-itiro Tomonaga, had devised a working example of a "renormalized" relativistic quantum-field theory which described the interactions of just two quantum particles, the electron and the photon; it was called quantum electrodynamics. Theorists focused their efforts on quantum electrodynamics not only because there were puzzling experimental data on the interactions of photons and electrons that required an explanation but also because photons and electrons, to a good approximation, were a little subsystem of all the quantum particles unto themselves. One could therefore ignore their interactions with other quantum particles—a vast simplification. If there was any validity to the renormalization procedure, it ought to work here.

When the renormalization procedure was carefully carried out, the calculational results of quantum electrodynamics could be compared with precision experiments. To many people's amazement, the theory, in spite of its abstract mathematical tricks, agreed decimal place for decimal place with experiments. Not since the time of Newton's predictions of planetary motions had theory and observation accorded with each other so completely. Even physicists were astonished by the experimental success of quantum electrodynamics.

After the success of quantum electrodynamics, physicists contrived to deepen their understanding of the renormalization procedure so that it seemed less a mathematical trick and more a profound feature of quantum-particle interactions. A major step was taken by Kenneth Wilson of Cornell University in the late 1960s. His work implied that in renormalizable theories the value of a quantum particle's mass or charge depended upon the distance scale at which we examined the particle. Viewed from a long distance, as it generally is, a particle has a definite mass. Viewed at microscopic distances, as is done in a high-energy accelerator, a particle may have an effective mass either larger or smaller than its long-distance value. This seems odd. How can the mass of a particle depend on the distance scale at which it is viewed? Usually we think of mass as something fixed and definite.

Imagine a straight line segment 6 inches long drawn on a piece of paper. That too seems like something fixed and definite. If we view the line from some distance, it appears shorter. Halve the distance, and it appears twice as long. Of course we are not fooled by this growing line segment—the original line is still 6 inches long. In fact, using our knowledge of our distance from the paper—our distance scale—and the angle subtended by the line we can easily calculate its length.

But now suppose we halved our distance from the line segment and instead of growing by a factor of 2 it grew by 1½ or even 2½. What calculation do we do then—what is the "true length" of the line segment?

Of course, line segments do not do that. But suppose instead of a line segment we take a picture of a coastline—a very twisted line— from on high in a satellite and measure its length between two points. Then we halve that distance and take another picture, measuring the length between the same points. One might think, in analogy with the line segment, that this length will double. But remarkably, it does not: it more than doubles. If we halve the distance scale once again, we find the same proportional amount of excess over the expected doubling.

We can mathematically describe such deviant scaling behavior by what the mathematician Benoit B. Mandelbrot calls "fractals" and physicists call "anomalous dimensions." Fractals, or anomalous dimensions, are just numbers that precisely specify, in any given example, the deviation from the expected scaling rule. Mandelbrot has found many examples of this odd scaling behavior in the natural world—it is often the rule rather than the exception. And quantum particles, described by renormalizable interactions, conform to that rule too.

As one examines quantum particles, their mass and coupling strength (which measures their interaction with other particles) change according to the distance scale at which they are examined, just as in the case of the coastline. The Princeton physicist Curtis Callen and Kurt Symanzik of Hamburg University, Germany, derived a set of equations that described this anomalous dimension behavior for relativistic quantum-field theories in 1968. Their equa-

A castaway on a "Koch Island" is viewed from two different distance scales. The coastline of a Koch Island exhibits "fractal," or "anomalous dimension," behavior—it does not change its apparent length in proportion to the distance from which it is seen. Just like a real coastline, the distance between the same two points increases more than the proportion expected as one moves closer

tions were based on Wilson's ideas and on earlier work by the physicists Murray Gell-Mann, Francis Low and A. Petermann. These mathematical developments did much to buttress the belief of physicists that the renormalization procedure was more than a mathematical trick—it had physical content.

Not every relativistic quantum-field theory is renormalizable—the mathematics of renormalization works for only a few kinds of quantum-particle interactions out of a possible infinite number. Remarkably, the renormalizable interactions are precisely the ones we observe. Is nature trying to tell us something by using only renormalizable interactions? Some physicists, struck by this fact, think that renormalizability is a fundamental imposition by nature, just like the principle of special relativity. Others are not so sure. But this much is clear: nature, by choosing renormalizable interactions among the quanta, has been kind to the theoretical physicists. Now, in principle, they can calculate the interactions of the quantum particles without getting nonsense for an answer.

GAUGE FIELDS

With the emergence of quantum theory as the language of nature, symmetry and group theory came to play an ever-increasing role in physics. Yet the most profound use of symmetry was not discovered until 1954, and its application to physics not realized until 1968. This discovery was the "non-Abelian* gauge-field theory" invented by the mathematical physicists C. N. Yang and Robert Mills.

Their basic idea was to generalize the notion of an internal symmetry. Suppose we have a three-component field, so that to the two components we previously called red and blue we add a third called

---

* "Abelian" symmetry operations, named for the Norwegian mathematician Niels Henrik Abel, obey the commutative rule $R_1 \times R_2 = R_2 \times R_1$, while the more general "non-Abelian" symmetry operations do not: $R_1 \times R_2 \neq R_2 \times R_1$.

---

to it—it effectively grows longer. The properties of quantum particles, like their masses and their interaction coupling strengths, also depend on the distance scales at which they are measured. Theoretical physicists conjecture that many interaction strengths become equal at very short distance scales, thereby realizing a unification of the forces of nature.

"yellow." We can imagine that red, blue and yellow correspond to three axes in a three-dimensional "internal space." The internal-symmetry operation would correspond to making an arbitrary rotation in this internal three-dimensional space of the field components. If we mathematically rotate the axes in this internal space, then the red, blue and yellow components of the field in real space are rotated to the same degree. If, when we do this, the total field energy remains unchanged, then a symmetry is present. In this case we speak of a "global internal symmetry" because the different components of the field have been rotated to the same degree over *all* of physical space.

Now imagine, as did Yang and Mills, that instead of rotating the field components to the same degree over all of space, we let the rotation of field components vary from point to point in physical space. This is called a "local internal-symmetry" operation because it differs locally, from point to point, and is not the same over all of space. But upon doing this we find that the total energy of the field is changed so that the original symmetry is now lost.

Yang and Mills discovered that the lost symmetry could be remarkably restored if one introduced yet another multicomponent field, called the non-Abelian gauge field, into real space. Allowing this additional multicomponent field to also rotate its many components into one another from point to point in real space restores the lost symmetry. The role of the gauge field is that it compensates for the loss of symmetry when we make the global internal rotation into a local rotation. We see that requiring the existence of a local internal symmetry—a rotation among field components that is allowed to change from point to point in physical space—has as a consequence a new field—the gauge field. The existence of gauge fields thus could be deduced from symmetry requirements alone. From this dramatic conclusion, placing the concept of symmetry even prior to that of a field, follows most of the contemporary research in relativistic quantum-field theory.

A way to visualize the effect of the Yang-Mills gauge field is to imagine a triangle on a spatial grid (see illustration). The triangle symbolizes the original multicomponent field, and the grid is a coordinate system that can represent the rotations in the internal space. A global rotation of the grid coordinate system does not alter the physical shape of the triangle—the physical situation remains unchanged. However, if the degree of grid rotation is varied locally

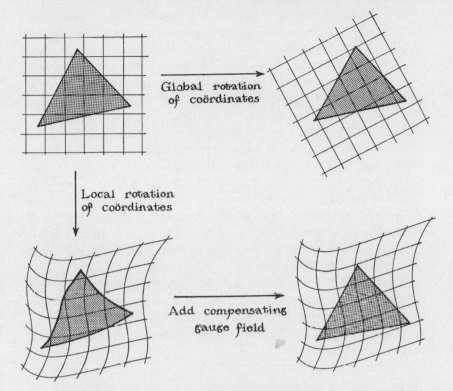

The triangle symbolizes a multicomponent field and the grid a coordinate system that can represent the rotations in the internal space. A global rotation of the grid leaves the triangle unchanged. However, a local rotation that changes from point to point distorts the triangle and so the physical situation is changed. When a Yang-Mills gauge field is introduced, the lost symmetry is restored.

from point to point, the shape of the triangle is changed and the symmetry is lost. The effect of the gauge field is to restore the lost symmetry so that the shape of the triangle is unaltered even when one distorts the coordinate grid at each point differently. Conversely, if we require that the shape of the triangle remain unchanged under arbitrary distortions of the coordinate grid, then we are required to introduce a compensating gauge field to restore the symmetry.

When Yang and Mills wrote their paper in 1954, it received little attention. Physicists admired the beautiful role of symmetry concepts it embodied, but did not see how these ideas could be applied to the problems with which they were then struggling—the problems of making realistic theories of the strong nuclear-strength interaction and the weak interaction. Two main theoretical obstacles stood in the way of application of the gauge-field concept to quantum-particle physics. The first was the problem of renormalizability—the non-Abelian gauge-field theory did not lend itself to the renormalization procedure that worked so well in the case of quantum electrodynamics. This problem was overcome only in the early 1970s when theoretical physicists, using a few new tricks, proved that the Yang-Mills gauge-field theory was also renormalizable.

The other problem was that nowhere in nature was the Yang-Mills type of symmetry apparent. Theoretical physicists believed that if the Yang-Mills symmetry was exact, then the corresponding field quanta, the particles, had to be exactly massless. None of the experimentally observed particles seemed to possess the properties required of massless Yang-Mills quanta.

Today we know the reason for this. The Yang-Mills field symmetries do not directly appear in nature. Instead, they appear indirectly in two ways: they can be exact but completely hidden symmetries, or they can be broken symmetries. Let us examine them in turn.

GAUGE SYMMETRY AND SYMMETRY BREAKING

Theoretical physicists have recently shown (by computer simulations of the field theory) that if the Yang-Mills symmetry is exact, then the symmetry remains completely hidden—all the components of the field that transform under the symmetry operation (like the red, blue and yellow components) have their associated quantum particles confined to a tiny region of space, and they never appear as true particles. They stay bound up and form a ball or bag—a massive particle. As I will discuss in the next chapter, such objects do exist; they correspond to the observed hadrons, the strongly interacting particles like the proton and neutron. At any rate, exact Yang-Mills symmetry implies confinement of the associated-field quanta, and that is why they do not appear directly in nature.

The second possibility for the Yang-Mills field is that the symmetry is broken spontaneously—the field equations possess the symmetry but the solution to the equations does not. Since it is the solutions to equations that describe the real world of quantum particles, one concludes that in the real world the original symmetry is broken and that is why we do not see it. But how can a symmetry just break like that?

Abdus Salam, the Pakistani physicist, gives the following example. Suppose people are invited to dinner at a circular table and precisely between each dinner plate and the next is a salad plate. The salad plates are symmetrically located between the dinner plates. The first person to sit down, not knowing the dinner rules, could as easily pick the salad plate to his right as the one to his left, and once he makes his choice the original symmetry is broken. Other people will have to follow suit—otherwise someone will be without a salad. It does not matter which choice is made—right or left: either choice breaks the original right–left symmetry. The solution to a symmetrical configuration breaks the symmetry.

Another example of a spontaneously broken symmetry, and one that is closer to real physics, is the "Heisenberg ferromagnet." A magnet consists of lots of little magnetic domains, which for our purposes we can imagine to be like little compass needles—small bar magnets free to pivot about. Suppose we lay thousands of such compass needles on a tabletop, each free to move. Imagine, too, that the table is shielded from the earth's magnetic field, so that the only magnetic field a compass needle responds to is the one produced by its neighbors on the table.

At first all the needles point in random directions. The net field produced by all the randomly oriented little magnets is on the average zero because the fields subtract as often as they add. Because there is no net magnetic field, were we to rotate in the plane of the table we would find no preferred north–south direction. The physical situation is thus rotationally invariant, or symmetrical, in the plane of the table.

Now suppose we manage to orient a bunch of the compass needles in one region so that they all point the same way, producing their own net magnetic field. We can do this by introducing a strong external magnetic field in that region and then shutting it off. The net magnetic field of all those oriented needles will soon cause all the other needles to follow suit and point in the same direction. The original rotational symmetry is now broken because there is a pre-

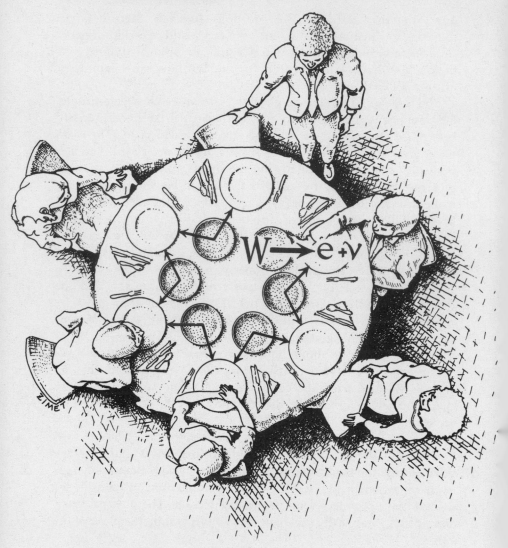

An illustration of a spontaneously broken symmetry—here the right–left symmetry of the salad plates between the dinner plates. If one person chooses a salad plate, the right–left symmetry is "spontaneously broken."

ferred north–south direction—the direction of the net magnetic field. Furthermore, this new configuration of all the little needles—this broken rotational symmetry—is clearly the stable one. If we manually change the orientation of one or two needles, they will spring back once released to their original orientation. The Heisenberg ferromagnet illustrates the basic ideas of spontaneous symmetry breaking: though the original physical situation is symmetrical, it is unstable; the broken-symmetry situation is stable.

The first suggestion that gauge symmetries could break spontaneously came out of the work of Peter Higgs, a British physicist, and Richard Brout and P. Englert, physicists at the University of Brussels, back in 1965. Higgs and I were both at the University of North Carolina when he was doing this work, and I remember him complaining about the trouble he was having with another physicist who was the referee of his article and thought his work was wrong. Fortunately, Higgs prevailed and the article, although delayed, was published. I also remember that at that time I could not see how his work pertained to real physics. It seemed like a nice piece of mathematics, a curiosity. Higgs himself was not sure what it was good for—but then, neither did most physicists see its application to real physics.

Then in 1967–1968, Steven Weinberg and Abdus Salam used Higgs's idea in a Yang-Mills gauge-field-theory model, which for the first time unified two distinct forces among the quantum particles —the electromagnetic force (describing the interactions of photons with matter) and the weak force (responsible for the decay of quantum particles). The electro-weak model of Weinberg and Salam incorporated the ideas of many other physicists, foremost among them Julian Schwinger, Sheldon Glashow and John Ward. Today physicists believe the model describes the real world. But this work was mostly ignored until 1971, when it was shown that the Yang-Mills–type theories were renormalizable. Then physicists could use the model to do detailed calculations of the weak and electromagnetic interactions just as they had done when quantum electrodynamics was invented. A revolution began in theoretical physics—the gauge-field-theory revolution—which continues to this day.

Higgs's idea was to introduce a new field in addition to the gauge field which is today called the "Higgs field"; it is spinless and has a mass. The virtue of the Higgs field is that physicists can use it mathematically to study the process of symmetry breaking in great detail. In a sense, the Higgs field is the "symmetry breaker"—the first person to pick a salad plate or the external magnetic field that forces the magnet needles to pick a common direction. By appropriately introducing the Higgs field one can mathematically show that the symmetry-preserving solution to the field equations is unstable—the symmetry "wants" to break, just as the magnet needles all want to point in the same direction. The unstable solution is like balancing a pencil on its tip—it is cylindrically symmetrical about its point, but unstable. A slight push will knock it over into an asymmetrical but stable configuration. The Higgs field, like the pencil, picks the stable but broken symmetry solution.

The symmetry breaking in the Higgs field affects the Yang–Mills gauge fields by breaking their perfect symmetry as well. The Yang–Mills fields in the symmetrical situation are all exactly massless, but when the gauge symmetry breaks, some of these previously massless gauge fields acquire a mass.

In the case of the electro-weak model, such massive gauge-field quanta correspond to the W and Z particles experimentally discovered in 1983 at CERN, a European high-energy laboratory. They have huge masses, more than 90 times the proton's mass—a consequence of broken symmetry. Remarkably, the observed masses of the W and Z particles accorded with the predictions of the theory, giving the theorists a big boost in confidence. Rarely in recent times have particle theorists had the pleasure of seeing abstract mathematical ideas realized in nature with such beautiful perfection. The idea of broken gauge symmetry seems here to stay.

Every success in physics creates new problems and puzzles on a deeper level. The deep puzzle is gravity. We have already seen how relativistic quantum-field theory is the offspring of the marriage of special relativity and quantum theory. But if we are to have a theory which includes gravity we must invent one that marries general relativity with quantum theory. In spite of the fact that some of the best minds in physics have struggled with this problem for decades, no one has succeeded in doing this in a consistent and coherent way. Though insights have been gained, a quantum theory of gravity eludes our grasp. Evidently new and profound principles are required before physicists can incorporate gravity into quantum theory.

Relativistic quantum-field theory is the abstract mathematical language of the quantum particles. With the success of the renormalization procedure and the gauge-field-theory revolution, most of the puzzles confronting particle physicists in the 1960s were solved in the 1970s. In the next chapter I will describe the "standard model"—

the current consensus on the field theories that describe the real world of quarks, leptons and gluons, the quanta out of which everything is made. The very success of these ideas has physicists excited even as they turn to ask: "What next?"

3

The Standard Model

Furthermore, in the search for new laws, you always have the psycholog-
ical excitement of feeling that . . . nobody has yet thought of the crazy
possibility you are looking at right now.
— Richard P. Feynman, Nobel Lecture, 1965

Occasionally, after years of experimentation and intellectual struggle
a new, coherent picture of the physical world emerges out of previous confusion. Such was the case with the invention in the late
1920s of quantum mechanics, which finally elucidated the weird
world of the atom that had baffled physicists for decades. A more
recent example is the gauge-field revolution, which resulted in the
late 1970s in the invention of relativistic quantum-field theories of the
strong, weak and electromagnetic forces. These theories, sought for
decades, brought order to the subnuclear world. The result of such
revolutions is often the establishment of a new scientific consensus, a
shared outlook about the world order. It is important for science to
have such an established consensus. It provides both a definite target
for critics and a firm ground from which adherents can launch into
flights of speculation.

Today the "standard model" of the interactions of subnuclear particles represents such a consensus. It has been experimentally successful—no experiment is inconsistent with the standard model. Some
physicists think it is still inadequately tested, but most think it is
correct. The model is a relativistic quantum-field theory in which the

quanta are called quarks, leptons and gluons—a set of fundamental particles. I will be describing it in detail in this chapter, but for now here is the basic idea in a nutshell.

Physicists have identified four fundamental interactions in nature: the strong nuclear interaction, the weak interactions that cause atomic nuclei and quantum particles to decay, the electromagnetic force and gravity. The standard model deals with three of these four forces, the strong, weak and electromagnetic forces. (Gravity, by far the weakest force, involves the so-far-unsolved problem of a quantum theory of gravity and is explicitly excluded.) In the standard model, each of these three forces is mediated by a set of quantum particles called gluons, which are quanta of a Yang-Mills gauge field. The strong force is mediated by a set of eight "colored gluons," the weak force by a set of "weak gluons" called the W and Z, and the electromagnetic force by the photon, the particle of light—also a gluon. All these gluons interact with a set of particles called quarks and leptons. The leptons are distinguished by the property that they interact only with the weak gluons and the photon, not with the colored gluons of the strong interactions. Quarks interact with all three sets of gluons but predominantly with the strongly interacting colored gluons. Gluons, as the name implies, cause the quarks and leptons to "stick together." Without gluons, the universe becomes unglued. It would consist of a gas of noninteracting quarks and leptons and not be very interesting.

The standard model neatly integrates two relativistic quantum-

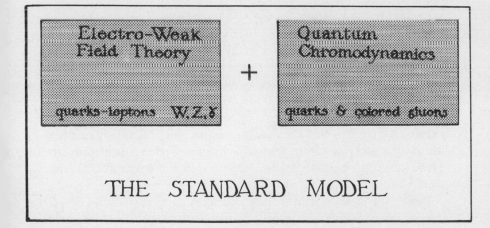

Electro-Weak
Field Theory

quarks-leptons W, Z, γ

+

Quantum
Chromodynamics

quarks & colored gluons

THE STANDARD MODEL

field theories—on the one hand quantum chromodynamics, a field theory of quarks interacting with the strong colored gluons, and on the other hand the Weinberg-Salam model of the unified weak and electromagnetic interactions. Joined together, these field theories can, in principle, account for everything we observe in the material world, save gravity.

Quantum chromodynamics mathematically describes how quarks bind together so tightly that they become permanently confined in tiny "bags." These baglike objects with the quarks trapped inside are the hadrons, the great zoo of strongly interacting particles observed in the high-energy accelerator laboratories. Included among these hadrons are the proton and the neutron, distinguished from the other hadrons by their relative stability. Protons and neutrons bind together to build up all the atomic nuclei. In a sense, nuclei are systems of quarks and colored gluons.

The electro-weak model unifies the previous theory of photons and electrons, called quantum electrodynamics, with a Yang-Mills theory of the purely weak interactions, which describes the decay of quarks and leptons. It was the first example of a unified-field theory in which two distinct interactions, in this case the electromagnetic and weak interactions, became but separate manifestations of underlying field symmetries. This electro-weak model has served as the inspiration for subsequent attempts at field unification.

The stable charged lepton is the electron, and this can combine with the nuclei, built up out of protons and neutrons, to make atoms. Atoms can make stars, planets, molecules and life. The standard model is the first step in the recipe for cooking up a universe.

In the previous chapter I described relativistic quantum-field theory, a conceptual framework for thinking about the microscopic world. Field theory provides a general language for discussing the quantum particles, the syntax or rules that any such description must satisfy. But it is another matter to discover the specific words—the quantum particles that appear in nature—which give real content to that language. Physicists, in their experimental and theoretical explorations, have uncovered the quarks, leptons and gluons—apparently irreducible units of matter out of which everything can be made. These quantum particles fit like words into the conceptual framework of the language of field theory and obey its rules.

Before we look at each of these quantum particles in more detail, it is worth reminding ourselves that in spite of the abstract language physicists use to describe them, these particles actually exist. No one else knows them as intimately as the experimental physicists who study them daily in the laboratory. The following story illustrates the difference between the way theorists and experimentalists view quantum particles: Arthur Eddington, the theoretical astrophysicist, and Ernest Rutherford, the experimental physicist, were having a dinner conversation back in the 1920s. Eddington remarked that atoms and electrons were perhaps just concepts. Thereupon Rutherford leaped up from the table as if the woman he loved had been insulted. Taking Eddington to task, he said that atoms were not just concepts; he met them every day in the laboratory and they were his friends.

LEPTONS

Physicists have evidence for a total of six leptons—the word means "light" and "swift." The leptons, consisting of the electron, muon, tauon and their neutrinos, can be thought of as tiny point particles without structure (at least, no structure has ever been seen). In the Wigner classification they all have a spin of one-half. Three of the six leptons are massive and have an electric charge of minus one—the electron, muon and tauon. Three are uncharged, and these are the three neutrinos (it is an open question whether they have masses). The six leptons can be divided into three "families" with a pair of leptons in each family. One member of each family has electric charge, while the other is the electrically neutral neutrino.

The first family of leptons, the "electron family," consists of the electron, denoted e, and its associated neutrino, denoted v_e. According to the rules of relativistic quantum-field theory there must be an antiparticle for every particle. The antielectron, or positron, is denoted $\bar{e}$, and the antielectron neutrino, a distinct particle from the electron neutrino, is denoted $\bar{v}_e$.

The electron, the first elementary particle, was discovered long ago in 1897. It is, as far as anyone can tell, absolutely stable and doesn't decay into other, lighter particles. The absolute stability of the electron is guaranteed by the law of electric-charge conservation—the total electric charge in a particle interaction must remain the same. The electron is the lightest charged quantum particle, and it cannot decay into lighter particles because there is no particle to carry away its electric charge. Like all physical laws, the law of electric-charge conservation is subject to experimental test, but thus far no one has seen the law violated.

Conservation laws, like the law of electric-charge conservation, play an important role in physicists' understanding of particle interactions and, as we will see, in their understanding of the origin of the universe. Conservation laws, according to Noether's work, are a consequence of exact symmetry. The electric-charge-conservation law, if absolute, is the result of an exact symmetry of the equations of field theory. We will encounter other such charge-conservation laws in my description of the particles, and all these conservation laws are consequences of internal symmetries of the standard model.

Electrons are perhaps the most familiar of all elementary particles because of their use in electronic instruments. An electric current is simply the movement of electrons or other charged particles. The swarm of electrons surrounding an atomic nucleus is found to be responsible for the chemical properties of atoms. Because electrons are so plentiful in nature and are easily liberated from their bonds to atoms, they have been extensively studied and their properties precisely determined.

The electron neutrino, the electron's family partner, is extremely light—it may have exactly zero mass. Experimentalists are trying hard to set limits on its mass or measure it if it is not zero. Neutrinos, because they have no electric charge, do not interact directly with the electromagnetic field. They have only very weak interactions with other matter and consequently fly right through us, the earth and anything else that gets in their way. Yet remarkably, experimental physicists, using extremely sensitive detectors, can identify subatomic events induced by neutrinos; they have even made beams of the elusive neutrinos.

The next family of leptons, the "muon family," consists of the muon, denoted μ, and the muon neutrino, denoted ν_μ. The muon is a particle which as far as anyone can determine is identical to the electron except that it is 207 times more massive. It has the same electronic charge and spin of one-half as the electron. The muon's family partner, the muon neutrino, may have no mass at all. The muon and its neutrino, like the electron family, have antiparticle partners. Muons, however, are unstable and don't hang around like electrons. That is because when they do decay into electrons, antielectron neutrinos and muon neutrinos, they can pass their electric charge on to the electron.

The decay of the muon exhibits yet other conservation laws, similar to the electric-charge conservation law—the laws of electron and muon number conservation. Suppose we assign a "muon number charge" of +1 to the muon and its neutrino and −1 to their corresponding antiparticles and zero for all other particles. Likewise we assign an "electron-number charge" of +1 to the electron and its neutrino and −1 to their antiparticles and zero to all other particles. Then we notice that the decay of the muon conserves both muon number and electron number. The charge-conservation accounting can be tabulated as follows:

$$\mu \rightarrow e + \nu_e + \nu_\mu$$

	$\mu \rightarrow$	e +	ν_e +	ν_μ
ELECTRIC CHARGE	−1	−1	0	0
MUON NUMBER	1	0	0	1
ELECTRON NUMBER	0	1	−1	0

Of course this is only one decay process, but when physicists look at lots of different interactions of muons and electrons they find that the new conservation laws all remain valid. Some physicists think that unlike the charge-conservation law, these new conservation laws will be violated, and that, though a violating interaction has a very low probability of occurring, it will someday be seen. These conservation laws correspond to symmetries in the standard model, so the model would have to be modified if the laws are violated.

As recently as 1977 yet another family of leptons was discovered —the tauon family. The electrically charged member is called the tauon; it is denoted τ and its associated neutrino ν_τ. The tauon is 3,491 times as massive as an electron. If the muon is a "heavy" electron then the tauon is a heavy muon. The tauon, like the muon, is unstable and decays into many other possible particles. But a "tauon number" is evidently conserved in all these processes. Tauon number, muon number and electron number are all conserved charges in the standard model. The sum of these charges is called the "lepton number," and since each individual charge is conserved, so is the sum.

We can summarize our classification of the leptons, our first set of blocks for building a universe, by arranging them in a table. With the three families denoted by I, II and III, the leptons fall into the follow-

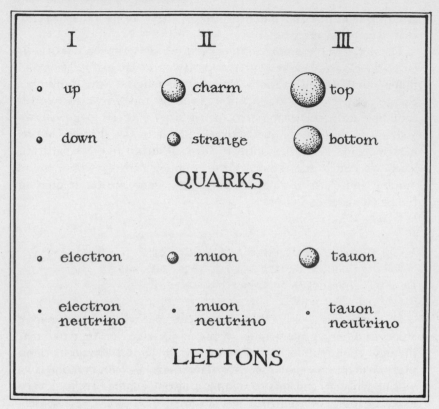

The detected quarks and leptons fall into three families. Quarks and leptons appear to be fundamental pointlike particles. The leptons, like the electron, can be directly detected, while the quarks exist only bound up together in particles called hadrons.

ing scheme, where next to the symbol denoting the particle we indicate its mass in units of the electron's mass.

ELECTRIC CHARGE	FAMILY I		FAMILY II		FAMILY III	
−1	e	1	μ	207	τ	3,491
0	ν_e	less than 0.00012	ν_μ	less than 1.1	ν_τ	less than 500

LEPTON TABLE

QUARKS

Leptons can be directly created and detected in the laboratory. Experimental physicists make beams of electrons, muons and even neutrinos and scatter them from other, target particles. If we now turn to the quarks, another set of point quantum particles of spin one-half, we find that in many ways they resemble the leptons. They too seem to be point particles without structure. But there are significant differences in how quarks and leptons appear in nature.

First, in contrast to leptons, no one has made beams of quarks, let alone ever seen a free single quark flying about. Quarks exist, but they always seem to bind together with other quarks so tightly that one cannot pull them apart. All evidence supports the conclusion that quarks are permanently confined in bound systems, little bags which can be identified with the hadrons—strongly interacting particles which we *can* observe in the laboratory. It was by using high-energy accelerators like microscopes and looking inside such hadrons as the proton and neutron that physicists first detected the presence of quarks.

There are lots of hadrons—an infinite number—but remarkably, they could all be built out of just a few quarks orbiting about each other in the bag in an infinity of different orbital configurations. The "standard model" maintains that there are six quarks (five have been experimentally detected) and that these quarks, unlike the leptons, have unusual fractional electric charges of $-\frac{1}{3}$ and $\frac{2}{3}$ times the unit charge. These six quarks also seem to organize themselves into three families consisting of a pair of quarks each.

The six quarks have been given names and symbols as follows:

    u − up quark
    d = down quark
    c = charmed quark
    s = strange quark
    b = bottom quark (sometimes called beauty)
    t = top quark (sometimes called truth)

The corresponding antiquarks are denoted $\bar{u}$, $\bar{d}$, $\bar{c}$ and so on. They are arranged into families I, II and III as follows:

ELECTRIC CHARGE	FAMILY I	FAMILY II	FAMILY III
⅔	u 2	c 1,500	t ?
−⅓	d 6	s 200	b 3,000

QUARK TABLE

All the quarks possess mass. One cannot, of course, directly measure the mass of something that is permanently trapped, but theoretical physicists can assign masses to the quarks on the basis of their properties even if bound up. The up quark is the lightest, with a mass approximately 2 times that of an electron. The other light quark, the down quark, has a mass about 6 times that of an electron, while the four other quarks are all quite a bit more massive. The approximate mass value is denoted in the quark table next to the particle's symbol (except for the top quark: although there is evidence that it has been detected, its mass is not known).

These quarks do not appear directly in nature but instead combine to form the hadrons that can be seen. The rules for building the observed hadrons out of quarks are rather simple and present what is called the "quark model of the hadrons." It was invented in 1963 by Murray Gell-Mann and independently by George Zweig in order to understand the systematic relations that were observed among the hadrons. Here are the quark-model rules.

Recall that all quantum particles have either integer or half-integer spin. Hadrons with a spin of half an integer are called "baryons"; those with a spin equal to an integer are called "mesons." If we denote any one of the quarks u, d, s, c . . . by the generic symbol q, then the baryons are made of three quarks:

$$q\,q\,q$$

while antibaryons are made of three antiquarks:

$$\bar{q}\,\bar{q}\,\bar{q}$$

The other great subdivision of the hadrons, the integer-spin mesons, are made out of a quark and an antiquark according to

$$\bar{q}\,q$$

With these elementary rules one can check that all the baryons and mesons, given the fractionally charged quarks of ⅔ and −⅓ and antiquarks of charge −⅔ and ⅓, have integer-valued charges of 0, ±1, ±2—just the charges observed in the lab. For example, the proton is made of three quarks according to

$$\text{proton} \sim \text{uud}$$

while the neutron is made of three quarks according to

$$\text{neutron} \sim \text{udd}$$

Since the u quark has charge ⅔ and the d quark has electric charge −⅓, one sees that the proton's charge is ⅔ + ⅔ − ⅓ = +1 and the neutron's charge is ⅔ − ⅓ − ⅓ = 0—just the right charge.

Other hadrons, the so-called "strange" particles, can also be built out of quarks if we replace a "d" down quark with an "s" strange quark. For example, the lambda, a "strange" particle observed in the lab, is constructed out of quarks by replacement of one of the d quarks in the neutron with an s quark:

$$\text{lambda} \sim \text{usd}$$

Likewise all the observed hadrons can be built up out of quarks.

The quarks inside hadrons can orbit about each other in an infinite variety of discrete configurations, and each of these configurations corresponds to another hadron. Usually only the lowest-energy orbital configurations, such as the proton, neutron or lambda, are observed in the laboratory. The higher-energy configurations are very unstable and decay rapidly into the lower-energy ones.

We see that quarks, although they cannot be directly detected, can be used to build up all the strongly interacting hadrons. Leptons, on the other hand, can be directly detected. But quarks and leptons, if we ignore differences in how they appear in nature, resemble each other—they are both point particles with a spin of one-half and seem to organize themselves into three families. This observation will be the springboard for speculative leaps that attempt to formulate a unified theory of the quarks and leptons.

But what about the interactions among the quarks and leptons? These interactions must be important, because they bind quarks together but not leptons. According to the "standard model," these interactions are mediated by the other set of quantum particles called gluons, the final set of quanta in the standard model.

GLUONS

Gluons are a new class of quantum particles with spin equal to one. Interestingly, from the point of view of relativistic quantum-field theory, gluons exist because of symmetry.

Recall that every quantum particle had an associated field. The fields associated with gluons are the Yang-Mills gauge fields. In the last chapter I described how the existence of Yang-Mills fields can be mathematically deduced if we postulate the existence of an "internal" symmetry—not just globally over the whole of space-time, but locally at each point in space-time. Requiring such a local internal symmetry turns out to imply the existence of a Yang-Mills gauge field, and the quanta of that field are the gluons. Gluons are therefore a consequence of symmetry.

The gluons, in their role as the mediators of interactions between quarks and leptons, can be thought of as quantum particles which are exchanged between two other quantum particles like a ball thrown between two ballplayers. Gluons have a characteristic coupling strength to the quarks and leptons which measures their stickiness— how strongly gluons latch on to the other particles. That coupling strength of the gluons is proportional to the different charges—generalizations of the idea of electric charge—that the quarks and leptons possess. For example, the photon, the particle of light, is a gluon, and it couples to the electric charge of other particles with a strength proportional to that charge. But there are other gluons which couple to yet other kinds of charge.

Physicists have learned that the gluons exchanged between quantum particles are responsible for all the forces of nature. Each of the three forces described by the standard model—the strong, electromagnetic and weak forces—has an associated set of gluons and a mathematical field theory which describes their interactions. The strong, quark-binding force is mediated by a set of eight "colored gluons" and is mathematically described by the field theory called quantum chromodynamics. The electromagnetic and weak forces are mediated by gluons known as the photon and the weak boson (denoted W and Z) and mathematically described by the electro-weak unified-field theory. Let us now have a more detailed look at these gluons and the field theories that describe their interactions.

COLORED GLUONS AND QUANTUM CHROMODYNAMICS

The quarks interact predominantly with a set of eight "colored" gluons. But what is "color"? Each quark is assumed to come in three charges—called "color" charge. Quarks aren't really colored, but it helps to imagine that each quark is either red, blue or yellow. For example, there are a red up quark, a blue up quark and a yellow up quark.

The role of the eight colored gluons is that they can exchange the color charges of the quarks. For example, if a red quark interacts with one of the eight colored gluons it can change into a blue quark. Not only do the eight colored gluons interact and exchange color charges among the quarks: they also interact among themselves, exchanging their color charges.

The field theory that mathematically describes these interactions among the colored quarks and eight colored gluons is called "quantum chromodynamics," or QCD for short. According to QCD, the colored gluons bind the quarks together into little bound systems that can be identified with the observed hadrons. Computer studies of QCD indicate that the binding between quarks due to the colored gluons is so strong that they never become unstuck; that is why quarks exist in a permanently bound state inside the baglike hadrons.

What happens if you try to liberate a quark out of its hadron prison? As you try to pull a quark from inside a hadron, you discover that the force of the colored gluons increases with distance. This means it gets harder and harder to pull them apart and you have to supply more and more energy to separate them. This energy becomes so great that in accordance with Einstein's mass–energy equivalence, the energy in the colored gluons binding the quark transforms into a massive quark and antiquark pair, each of which then becomes part of a hadron. Instead of liberating a free quark you end up creating two hadrons!

Besides offering an explanation for quark confinement, quantum chromodynamics also explains the quark model of the hadrons—it automatically gives the rules that tell us which combinations of quarks will stick together to make the hadrons. The basic idea embodied in the mathematics of QCD is that quarks and gluons, although they have colored charge, prefer to form combinations of themselves that are "color-neutral" and have no color charge. A

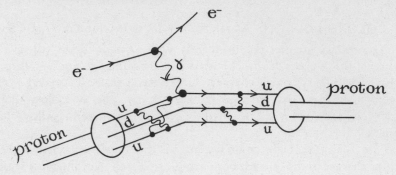

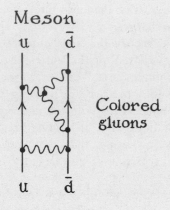

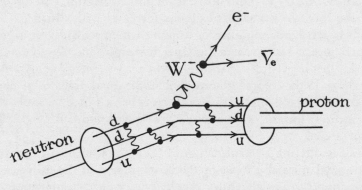

Diagrams of quantum-particle interactions according to the "standard model."
According to quantum chromodynamics, strongly interacting hadrons like the
proton and neutron are combinations of quarks which are bound together by

simple way of thinking about this is to imagine the three colors of a quark to be the three primary colors—red, blue and yellow—and the three antiquarks to have the three complementary colors. If you equally mix together the three primary colors (or their three complements) you get neutral white. So three different-colored quarks together or three antiquarks together form a "color-neutral" combination. But, three quarks together make a baryon and three antiquarks make an antibaryon. This is just the rule of the quark model, here obtained by the requirement that color is "confined"—only color-neutral combinations may exist. We go on to notice that a color and its complement form a gray mixture—again a color-neutral combination. This corresponds to combining a quark with an antiquark to make a meson. We see that the requirement of color confinement or neutrality supplies the rules for making hadrons. The quarks and gluons may be blazing in brilliant "colors," but they appear only in black-and-white combinations corresponding to the observed hadrons.

Quantum chromodynamics as a theory for the quark-binding force came into existence in the early 1970s, proposed independently by quite a number of theoretical physicists. These physicists knew about the quark model and the fact that the color-quark idea gave the right rules for building hadrons. What was previously lacking for the invention of QCD was proof that the Yang-Mills field theory of the colored gluons was renormalizable. Once mathematical physicists proved that field theories such as QCD were indeed renormalizable, the excitement began—QCD became a viable theory.

When theoretical physicists began to mathematically explore the renormalization properties of QCD, they made a remarkable and completely unanticipated discovery: at very high energy or corre-

---

color gluons, here indicated by the wavy lines connecting the quarks. The top diagram shows a electron (e⁻) scattering from a proton made out of quarks (uud) by exchanging a photon (γ)—a quantum of light, which is also a gluon. It was through processes of this kind that the presence of quarks was first detected in high-energy-physics labs. The middle diagram is a representation of a pi meson, another strongly interacting quantum particle, composed of a quark (u) and an antiquark (d̄), here shown exchanging colored gluons. The bottom diagram represents neutron decay into a proton, an electron (e⁻) and an antielectron neutrino (v̄ₑ). The decay occurs because a weak gluon (W⁻) can change a down quark (d) into an up quark (u). The colored gluons cannot do this.

spondingly short distances, the coupling strength of the colored gluons became weaker and weaker until at infinite energy it became zero—the interaction between the quarks and colored gluons vanished. This odd property, unique to field theories of the Yang-Mills type, is called "asymptotic freedom"—at asymptotically high energy the quarks and colored gluons behave as if they were free and noninteracting. This mathematical discovery was made by Hugh David Politzer at Harvard University and independently by David J. Gross and Frank Wilczek at Princeton University in 1973. It was based, in part, on the earlier work of Curtis Callen, Kurt Symanzik and Ken Wilson on renormalization theory. This discovery about QCD fitted in beautifully with experiments done previously at the Stanford Linear Accelerator Center and other high-energy physics laboratories which already showed that quarks, if viewed at short distances inside the proton and neutron, indeed behaved as if they were free particles inside their baglike prisons. These experiments lent support to the idea of asymptotic freedom and the growing conviction that QCD was the correct theory of the quark-binding interaction.

QCD has a number of internal symmetries which imply the existence of charge-conservation laws—laws that are manifested in the strong interactions of the hadrons. Color charge is a conserved quantity, but since all hadrons are color-neutral, there is no way to see this conservation law in action in the laboratory. It would be like postulating an electric-charge-conservation law for a world consisting of only electrically neutral particles—you would never see any electric charge to check if it was conserved.

But there are other charge-conservation laws that apply to hadrons which can be checked in the laboratory. The number of quarks of each kind—up, down, strange, charmed and so on—must be conserved (antiquarks are counted negatively). This means that in an interaction between the baglike hadrons, the number of up or down quarks remains the same. They can jump from hadron to hadron at the moment of collision when the bags overlap, but their total number does not change. Since antiparticles are counted negatively, the law of up-quark-number conservation also allows for the creation of an up quark and an anti–up quark out of pure energy. The various quark-number-conservation laws, when applied to the strong interactions of the observed hadrons, are confirmed in thousands of laboratory experiments—no one doubts them.

The weak interactions, which I have yet to discuss, violate these various quark-number-conservation laws. For example, a charmed quark can change into a down quark by the weak interaction, and this violates the conservation of charmed-quark number and down-quark number. But even the weak interaction preserves the law of total-quark-number conservation. The number of quarks minus the number of antiquarks is strictly conserved in the standard model.

This apparently absolute law of total-quark-number conservation implies a strict conservation law in the corresponding hadron interactions—the law of baryon-number conservation. Baryons are the family of half-integer-spin hadrons, and in any interaction this law implies that their total number must be conserved. The proton, since it is the lightest baryon, must be absolutely stable as a consequence of the law of baryon-number conservation. There is no particle to which it can pass down its baryonic charge, in the same way that the electron has no lighter particle to which it can pass its electric charge. A good thing, too! Protons make up most of the visible matter in the universe, and if they could decay quickly the universe would decompose. Later, in the light of other field theories that go beyond QCD and the standard model, we reexamine the law of baryon-number conservation and proton stability. These other field theories imply that the proton *is* unstable, though its rate of decay is so slow that the universe remains unaffected.

According to theorists, quantum chromodynamics offers a complete mathematical model of all the strong interactions of the hadrons. It fulfills the dream of decades to find an exact theory of the strong nuclear force. Unfortunately, although QCD is a mathematically elegant field theory based on beautiful symmetries, it is very difficult to solve its equations and extract precise results that can be compared with the results of experiments. But for those few details of strong interactions which *can* be extracted from the theory and compared with experiment, the agreement is fine. Further comparisons between theory and experiment will have to wait until theorists solve the equations of QCD on powerful computers—a research area of major activity. In spite of the current difficulties, most theoretical physicists are confident that they have found the right theory of the strong quark-binding force. Sometimes I like to tease my experimental colleagues by suggesting that their difficult strong-interaction experiments are no more than analogue computations for solving the equations of QCD.

ELECTRO-WEAK GLUONS AND THE WEINBERG-SALAM MODEL

The world of quarks and colored gluons, whose interactions are so strong we can virtually ignore everything else, constitutes an exclusive world of interactions in its own right. But there are other interactions—the weak and electromagnetic interactions—which involve both quarks and leptons. The electromagnetic and weak interactions used to be viewed as separate interactions. But today they are seen as a unified electro-weak interaction mediated by a set of four gluons consisting of the photon, denoted γ, which is the quantum of the electromagnetic field, and the three weak gluons consisting of two electrically charged gluons, denoted W^+ and W^-, and a neutral one, Z^0, which are quanta of a Yang-Mills field. These are the electroweak gluons, and they are responsible for new forces between leptons and quarks.

The photon, the first gluon to be directly detected as a quantum particle, couples to the electric charge of particles. Unlike the colored gluons, which produce a force between quarks that increases with distance, photons produce a force between particles that decreases with distance—the Coulomb force, described in the eighteenth century by the French physicist C. A. de Coulomb. Because of this, electrically charged particles like electrons can be separated from the rest of matter and produce long-range electric fields which stretch out over macroscopic distances. The same is true for magnetic fields. For example, the field of a bar magnet can stretch out over large distances—large as compared with atomic distances. Because of the long-range nature of electromagnetic fields, the fact that light is an electromagnetic wave and the important role of the electromagnetic field in binding electrons to nuclei, the electromagnetic interaction is easily studied in the laboratory. It was the first to be tamed by theorists and today is the best understood of all interactions.

The first modern relativistic quantum-field theory, called quantum electrodynamics, was devised back in the 1940s and summarizes everything we know about the interaction of light with electrons. Quantum electrodynamics proved so successful in accounting for the experimentally observed properties of the electromagnetic interaction that it became the exemplar of all future field theories.

What are the weak interactions? Unlike the electromagnetic interaction, the weak interaction is very short-range and makes itself observable only on the subnuclear distance scale through the decay of quantum particles. An elementary and closely studied example of the weak interaction is the decay of a free neutron into a proton, an electron and an antielectron neutrino, a decay that takes place, on the average, every 1,000 seconds. Almost all quantum particles eventually decay into lighter ones.

Weak interactions, in contrast to the electromagnetic interaction, were very complex and puzzling, and it took physicists decades to unravel their properties. Part of the difficulty lay in the extreme weakness of the weak force; another difficulty lay in the fact that it was extremely short-range. Yet were it not for the existence of the weak interaction known to be mediated by the massive weak gluons, denoted W^+, W^- and Z^0, the heavier quarks and leptons would be absolutely stable and not decay into lighter ones. Lots of exotic forms of matter could then exist indefinitely rather than existing for only brief instants in high-energy collisions between quantum particles. If the weak interaction that removes "strange" and "charmed" particles from the world by letting them decay were "shut off," the world would be very odd indeed. All those exotic particles could become chemical building blocks for new forms of "strange" matter.

The role of the weak-interaction gluons, the W^+, W^- and Z^0, is that they change quarks into other quarks and leptons into other leptons. A charmed quark through its interactions with a W gluon can be changed into a strange or a down quark. Likewise the weak gluons interact with the leptons. A tau lepton, through its interaction with a W gluon, can be changed into a tau neutrino. In general, the existence of these weak gluons means that all of the various quarks— up, down, strange, charmed and so on—can change into one another and different leptons within a family can change into one another. But quarks cannot change into leptons and vice versa, because these weak interactions respect various number-conservation laws.

The tendency in nature is for the heavier particles, since they have the most mass-energy, to release this energy by decaying into the lighter particles with the energy of the original particle transformed into the energy of motion of the lighter particles. Hence, the heavier strange or charmed hadrons—hadrons containing a strange or charmed quark—will decay into lighter hadrons not containing these massive quarks. Ultimately only the electron and the neutrino and the up quark, since it is the lightest-mass quark, are stable. Since quarks can appear only trapped inside baryons, this means that only the lightest baryon—the proton—is stable. The stability of these various particles is guaranteed by the charge-conservation laws and the fact that they are the lightest particles bearing a conserved charge. It is important that such stable particles exist. Otherwise we would have nothing out of which to build the visible universe.

The relativistic quantum-field theory that describes the unified weak and electromagnetic interactions was discovered by Steven Weinberg and Abdus Salam in 1967–1968. It not only correctly described the weak and electromagnetic interactions as they were then known but went on to predict completely new features of these interactions, features which, when they were subsequently observed experimentally, lent great credibility to the theory. Perhaps the most dramatic discovery was made in 1983 by experimentalists at CERN when they detected the W and Z gluons at the predicted mass values —a beautiful confirmation of the electro-weak theory.

All the previous gluons I've discussed—the colored strongly interacting gluons and the photon—have zero mass. By contrast, the weak gluons W and Z are very massive. How can such a big difference in mass between the W and Z gluons and the photon come about if the weak and electromagnetic interactions mediated by these very same gluons are truly unified? The answer to this question lies in the notion of "broken symmetry."

The electro-weak synthesis of Weinberg and Salam utilizes the idea of broken symmetry as its very essence. This idea can be explained rather simply. In the mathematical quantum-field theory these scientists devised, the photon and the weak gluons are all quanta associated with different components of Yang-Mills fields. These fields possess a Yang-Mills symmetry which transforms them into one another so that they are all manifestations of underlying unified fields, and in this sense they are related. But the electro-weak model, besides having Yang-Mills, lepton and quark fields, has a new ingredient, the Higgs field. This is the spin-zero massive field which plays a crucial role in determining how all these fields are eventually manifested in nature. The Higgs field also has several components, and if one writes down the equations one finds that they also have the Yang-Mills symmetry. But the stable solution to these equations for the Higgs field does not have symmetry; instead, it manifests a broken symmetry. Since the Higgs field interacts with the other fields, their equations, although symmetrical, also now have solutions in which the symmetry is broken. A symmetry once broken affects all the fields.

The Higgs field is a kind of "symmetry breaker" that destroys the original Yang-Mills symmetry. But in the case of the weak interactions, broken symmetry is just what is wanted to distinguish the electromagnetic from the weak interactions. The W and Z fields become very massive because of this symmetry-breaking "Higgs mechanism," while the photon, which retains a remnant of the original exact Yang-Mills symmetry, remains massless. Although the underlying theory is symmetrical and unified, its manifestation in the real world is not. As Weinberg commented, "Even if a theory postulates a high degree of symmetry, it is not necessary for . . . the states of the particles to exhibit the symmetry. . . . Nothing in physics seems so hopeful to me as the idea that it is possible for a theory to have a high degree of symmetry which is hidden from us in ordinary life."

The Higgs field seems to play a crucial role in breaking the symmetry of the equations, and the quantum particle associated with this field ought to be detected someday by the experimentalists. Unfortunately, unlike the prediction of the W and Z masses, the theory does not give a precise prediction for the mass of the Higgs particle. Yet it would be a powerful vindication of these ideas if a Higgs particle were produced and detected in the laboratory. Experimentalists at the high-energy lab at CERN are going to look for it now that the W and Z have been detected.

There you have it: the standard model consisting jointly of quantum chromodynamics for the strong interactions of the quarks and the electro-weak theory for the electromagnetic and weak interactions of the quarks and leptons. It seems consistent with all experiments; but more tests remain to be done. In 1984, experimentalists were reporting a few high-energy events seen at CERN that might be difficult to explain in terms of the standard model. But the meaning of these events is not clear. Most particle theorists take the standard model as gospel truth; it describes how to build a universe, and any future model must include it.

BEYOND THE STANDARD MODEL

Yet most physicists feel the standard model is ultimately unsatisfactory and that it cannot be the final word. They think we must go beyond it in order to have an even more fundamental understanding of the quantum particles. The basic reason for this dissatisfaction with the standard model, in spite of its successes, is that it leaves open a number of fundamental questions. As long as such questions can be asked, physicists know they have not finished their work.

The standard model takes account of nineteen parameters—masses of quarks and leptons, coupling strengths and so on which are input parameters that must be determined beforehand by experiments. Given those numbers, one can in principle mathematically describe all the strong, weak and electromagnetic interactions we observe in the laboratory. That is a great accomplishment. Yet most physicists feel that the ultimate theory should have no input parameters, no fundamental dimensionless constants, and that all the masses of quarks, and all coupling strengths, should be predicted by such a master theory.

Einstein wondered whether God had had any choice in creating the world the way He did and once wrote:

> Concerning such [dimensionless constants] I would like to state a theorem which at present cannot be based upon anything more than upon a faith in the simplicity, i.e., intelligibility of nature: there are no *arbitrary* constants of this kind; that is to say, nature is so constituted that it is possible logically to lay down such strongly determined laws that within these laws only rationally completely determined constants occur (not constants, therefore, whose numerical value could be changed without destroying the theory).

If Einstein is right and the master theory, yet to be discovered, allows for no free parameters, then God had no choice—He could not have "adjusted" the parameters, like the masses of quarks, to make different universes.

Another reason the standard model dissatisfied many physicists is that it is incomplete in two respects. First, it does not include gravity and the principles of the general theory of relativity. Second, the unification of the fields already in the standard model is still incomplete. Although the electromagnetic and weak fields are unified in the

Weinberg-Salam model, the strong colored field is not unified with either of these.

Some physicists, encouraged by the success of the electro-weak synthesis, have gone on to include the strong interaction in yet a further field unification and constructed models known as "grand unified theories," or GUTs for short. According to GUTs, the weak, electromagnetic and strong colored gluons are all quanta associated with a single multicomponent Yang-Mills gauge field. In the simplest such model, called the "SU(5) model," the single unified Yang-Mills field has 24 components, of which 12 correspond to the 3 weak gluons, 1 photon and 8 colored gluons of the standard model and 12 correspond to altogether new gluons. This multicomponent Yang-Mills field interacts with the quarks, leptons and also a new set of Higgs fields which break the single symmetry distinguishing the strong, weak and electromagnetic interactions.

According to this model, the 12 new gluons get an enormous mass when the symmetry is broken, a mass so enormous that these new particles will never be detected by any particle accelerator because no accelerator will ever have the energy to create them. These 12 new gluons have an important property that the strong, weak and electromagnetic gluons of the standard model do not have: they can change quarks into leptons and vice versa. These supermassive gluons have interactions that violate quark- and lepton-number conservation laws. This means that the interactions of these 12 supermassive gluons can cause one of the quarks inside the proton to change into a lepton. The proton, which was previously absolutely stable in the standard model, now becomes unstable in the simplest GUT model. Proton decay seems to be a natural consequence of the grand unification idea, although one can make special GUT models in which proton decay does not occur.

The proton-decay lifetime has been calculated in the simplest SU(5) GUT model to be about 10^{31} years—a very long time indeed: billions of times the present age of the universe. But already in 1982 experimentalists, searching for proton decay in gigantic detectors—swimming pools filled with water surrounded by photoelectric tubes that can see the products of proton decay—showed its lifetime to exceed 10^{31} years. The simplest SU(5) GUT model is therefore ruled out. The experimentalists are continuing to run their detectors in the hope of finding a decay. It would be very exciting to physicists if proton decay were eventually observed—it would support the GUT idea. By then studying the details of the proton-decay process, physicists might get an experimental handle on which particular GUT model was favored, if any.

Theoretical physicists have gone beyond the standard model in other ways, even beyond GUTs. One set of ideas goes by the name of "supersymmetry" and "supergravity." Supersymmetry is a new kind of symmetry that transforms fermions—fields with half-integer spins—into bosons—fields with integer spins—and vice versa. It offers the hope of still further unification. Supergravity is a local version of supersymmetry, just as Yang-Mills symmetry is a local version of ordinary "internal symmetry," and it incorporates gravity into the field-unification program. Supergravity models that unify all the interactions have been invented by the clever theorists, but they are not very realistic.

GUTs and supersymmetry are "wild ideas" at the edge of current theoretical-physics research. They may be right or wrong; there is no experimental evidence for or against them. But they hold out the hope of a new and more coherent account of nature and have attracted many bright young physicists to work on them. I will describe these efforts in more detail in a future chapter.

It is clear to physicists that many of these wild ideas will never be tested by conventional high-energy accelerators. The new theories describe regions of space and time so small and energies so high that no accelerator built on earth will ever probe them. The only time such energies were available was before the first billionth-billionth of a second in the big bang. Increasingly, theoretical physicists are turning to the dynamics of the early universe and looking there for clues that might rule out or support their wild ideas.

Before I report on these wild ideas, let us see what the better-tested standard model implies about the early universe. But first we must learn a little about thermodynamics and cosmology.

4
Thermodynamics and Cosmology

The second law of thermodynamics predicts that there can be but one end to the universe—a "heat death" in which [the] temperature is so low as to make life impossible.

—Sir James Jeans

In the last two chapters I described relativistic quantum-field theory —the language of quantum particles—and the contents of the microscopic quantum world—the "standard model" of quarks, leptons and gluons. With this information we are almost ready to descend backward in time to study the big bang. But before we do so, a few further concepts need to be described—thermodynamics and the role of the overall cosmological geometry of space and time. Then we will fit together these three ideas—the physics of quantum particles, cosmology and thermodynamics—and make a mathematical model of the early universe.

Our model for the early universe will be a gas of quantum particles uniformly filling the entire universe. The added feature that cosmology brings to this model is that space can contract or expand in time —a feature that influences the gas in that space. I have already described the quantum particles and their interactions, and the FRW cosmologies for the global structure of space and time. I will now discuss a few features of the thermodynamics of gases required to complete our model of the early universe.

For the moment forget about the universe and cosmology. Think of a gas trapped in a container with a definite temperature, pressure and volume—macroscopic properties which characterize the state of the gas. The physical laws of thermodynamics that relate these macroscopic properties of a gas to one another were already understood by nineteenth-century physicists. Yet it was not until physicists adopted a deeper viewpoint that the significance of these thermodynamic laws was finally recognized.

The deeper viewpoint is achieved if we remember that gases are not the continuous media they superficially appear to be, but in fact consist of huge numbers of particles bouncing around hitting each other or the wall of the container. Physicists mathematically derived the previous laws of thermodynamics by assuming that each particle obeys Newton's mechanical laws of motion and using an averaging procedure over the motion of all the particles. This new development, called "statistical mechanics," created a new and profound view of the nature of the collective properties of matter. For example, according to statistical mechanics the temperature of a gas is proportional to the average energy of motion of all the particles (the faster the particles move the higher the temperature) and its pressure is proportional to their average momentum. In this way the macroscopic variables describing a gas can be understood as measuring collective properties of all the gas particles.

Gases possess other macroscopic properties besides temperature and pressure. Among them is entropy—a statistical measure of the messiness of all the particles bouncing around. To illustrate entropy, suppose the gas container is filled with two different gases; call them A and B. We could imagine that in a starting configuration all the A particles are segreated in one half of the container and all the B particles in the other half, separated by a barrier. Then we remove the barrier. The A and B particles begin to mix, and soon the container is filled with a uniform mixture. How can we describe what has happened in terms of entropy?

Entropy is a measure of the degree of disorder of a physical system. But how does one measure disorder? The basic answer comes from probability theory, the mathematical study of randomness. Improbable configurations of all the gas particles are considered "ordered" and are assigned a low entropy, while probable configurations are the most "disordered" and have high entropy. For example, if we consider dealing card hands in a poker game, the likelihood is that most hands dealt are disordered sets of cards. These "messy" configurations have high entropy. The small number of desirable hands have a low probability of occurring—a set of card configurations with low entropy.

Now let us apply this thinking to our gas container. When the barrier is first removed, the A and B particles are still segregated—an improbable configuration corresponding to a relatively ordered state. The particles begin mixing because a mixed configuration of A and B particles is the more probable state. We notice that entropy, a measure of the degree of disorder, increases during the mixing. This increase in entropy—the change of a closed physical system from a configuration with a relatively low probability to one with higher probability—is called the second law of thermodynamics and is one of the cornerstones of statistical mechanics.

When gas like the one we are considering reaches a state of maximum entropy—that is, particles are thoroughly mixed and maximally messy—it is said to be in an "equilibrium state." There is nothing you can do to increase its messiness; hence it is in equilibrium, achieving the stability of complete disorder. Strictly speaking, we should refer to this as a state of "thermal equilibrium," meaning that the temperature is uniformly the same throughout the gas.

Gases in a state of thermal equilibrium have many important properties which can be rigorously proved by the mathematics of statistical mechanics. For example, one might think that the properties of such a gas depend on the details of the interactions between all the different gas particles, how they bounce off each other, how they hit the wall of the container. But remarkably, according to statistical mechanics, knowing those details is completely unimportant. All one needs to know to determine the physical state of the gas is the fact that different particles do interact and collide in some way so that they can transfer energy to one another.

What, then, is important for specifying the physical state of the gas? The amazing feature of a gas in thermal equilibrium is that once we know its temperature and the densities of the conserved quantities in the particle interactions, the state is specified. In the gas we have been considering, the number of A particles and the number of B particles are conserved quantities. To determine their densities, we just divide the total number of particles by the volume they occupy. Once we know these particle densities and the temperature, we know the state of the gas.

These same features apply to gases of the quantum particles. The quantities required to specify the state of the gas are the temperature, the number of various kinds of quantum particles in the gas and the density of conserved particles in the interactions—the electric charge, lepton number and baryon number. That is why those conservation laws will prove to be so useful when we apply statistical mechanics to the early universe.

Quantum particles, however, obey the laws of quantum mechanics, not Newtonian laws, and that modifies some of the equations of statistical mechanics. Physicists have worked out all those modifications so that statistical mechanics can be precisely applied to gases of quantum particles. But those changes will not affect the qualitative features of gases I have already described.

The entropy of a gas of particles in equilibrium is easy to calculate; according to statistical mechanics it is proportional to the total number of particles. The more particles in the gas, the messier it can become, the larger its entropy. If a gas consists of A and B particles, then one may consider separate entropy for the A and B particles because the number of A and B particles may differ. Then one refers to a "specific entropy," 'which is the ratio of the total entropy to that of the A particles or B particles.

So far I have been describing a gas in equilibrium at a fixed volume and at a fixed temperature. What happens if we now expand the volume? Imagine a piston moving out of the container, so that its volume expands. Furthermore, suppose we change the volume slowly as compared with the average collision time between the particles. This implies that the gas always remains in thermal equilibrium because the particles have enough time to transfer their energy to one another during the expansion. This slow expansion is called an "adiabatic expansion," and during it one can show that the entropy of the gas remains constant.

In summary, the fundamental ideas about gases are that first, in a situation of thermal equilibrium a gas is described by its temperature and the density of the various conserved quantities; and second, that in an adiabatic expansion or contraction the total entropy, proportional to the total number of particles, remains constant. These elementary properties of gases in thermal equilibrium may be applied to a description of a gas filling the whole universe. But the universe is not a simple container of gas—it has no edges or sharp boundaries. Furthermore, the nature of its volume expansion is different from a piston moving out of a container, and that makes for modifications which we must take into account.

Imagine that the entire universe is filled with a uniform gas of particles and that the space-time of the universe is one of the homogeneous, isotropic FRW models: spaces that contract or expand depending on whether we move forward or backward in time. We may apply the rules of thermodynamics and statistical mechanics to this gas filling the universe, provided we take note of an important difference between the universe and a container of gas. Unlike a container of gas, the universe has no boundary; it either is infinite or closes on itself. The universe expands because space itself is stretching—the Hubble flow—not because its boundary is moving like a piston in a container. If we made a colossal triangle with laser light beams and floated it in space, then as the universe ages that triangle would expand in space. Likewise the gas filling the universe is subject to this same Hubble flow.

Suppose we removed the walls from our gas container. Then the pressure of the gas on the walls would drop to zero and the gas would explode into the surrounding space. The gas of photons filling the universe also has a pressure, but there are no walls to contain it. What, then, causes the pressure? One might be tempted to think that this pressure is caused by the expansion of the universe. But that is incorrect. The expansion of the universe is the expansion of space itself, not the expansion of anything *in* the space of the universe. The photon gas moves *with* the general expansion of space—its motion is *not* analogous to the expansion of gas in a container. Photons can produce a pressure simply because they are particles with energy moving at the speed of light, flying every which way and hitting anything in their way. That bombardment of photons produces a radiation pressure.

Once we understand the proper application of thermodynamics to the whole universe, it becomes a powerful conceptual and calculational tool. Applying this thermodynamic viewpoint to the universe as it appears today, physicists simply approximate everything in the universe as a gas filling the universe. This gas consists of two important components.

The first component is matter: the galaxies, stars and any invisible dark matter—essentially a "gas" of massive objects which doesn't move very much. This matter gas is "cold"—it has zero temperature

—because temperature is a measure of the average energy of random motion.

The second component of the universe is radiation—the gas of microwave background photons detected by Penzias and Wilson.

Which of these two components, matter or radiation, dominates the mass density of the universe? The answer to this question is important, because according to Einstein's theory of gravity, it is the mass density of the universe that controls its expansion rate—the greater the mass density, the slower the expansion. If we measure the contribution of matter to the mass density of the universe today and compare it with the mass-energy density of the gas of photons, we find that the matter density is larger by at least a factor of 1,000 —the universe is "matter-dominated" not "radiation-dominated." We conclude that the gravitational dynamics of the whole universe today—its expansion—is controlled by its matter content, not its radiation content.

Although matter clearly dominates the energy density of the universe today, in the distant past, during the period of the big bang, the radiation was the dominant component and controlled the dynamics of the expansion. How do we know that? As we go back in time the size of the universe contracts, heating the gas of particles within it and causing the temperature to rise. The energy density of the matter increases, but the density of the radiation energy increases more rapidly and eventually overtakes the matter energy density. It is not difficult to explain why.

A photon, part of the background gas of radiation, is characterized by a wavelength inversely proportional to its energy. "Hot" photons are blue and have short wavelength, and "cool" ones are red with long wavelength. If one considers a gas of photons with lots of different wavelengths, then the temperature of the gas is the average energy of the photons in the gas. So we conclude that the average wavelength of a photon in the gas is inversely proportional to the temperature of the gas. If we imagine the universe contracting, then all the photons in the universe are blue-shifted—their wavelength decreases, their average radiation energy E_R increases and hence their temperature T increases proportionally: $E_R \sim T$.

If we next consider any volume in space, V, occupied by the gas of photons, it too contracts with the space. Since a volume is the cube of a length, and since all lengths, like the wavelengths of photons, are contracting with the inverse temperature, we conclude that any volume in space is decreasing as the inverse cube of the temperature: $V \sim T^{-3}$. The energy density of the photon gas is the energy of the photons divided by the volume of space they occupy. Since the average energy of the photons, E_R, is proportional to the temperature, and this is to be divided by the volume, V, we conclude that the energy density, E_R/V, of the photon gas is proportional to the fourth power of its temperature: $E_R/V \sim T^4$—a relation known after its discoverers as the Stefan-Boltzmann law. What this all comes down to is that if you know the temperature of a photon gas you also know its energy density. Since we know that for the universe today the temperature of this gas is about 3 Kelvin, we can thus calculate the radiation energy density and compare this with the matter energy density.

As I stated earlier, the matter density today may consist of contributions from both visible and invisible forms of matter. Just the visible-matter density estimated by astronomical observations is roughly 1,000 times the energy density in radiation as calculated by the Stefan-Boltzmann law. Including possible invisible matter only increases the matter density. This, then, is how we know the universe is matter-dominated today. But what about the past?

In order to compare the energy densities of matter and radiation in the past we also have to know how the matter density depends on temperature. The energy density of matter, since matter can be regarded as not moving and cold, is just the mass-energy E_m of the matter—a fixed quantity independent of temperature—divided by the volume it occupies, V. Hence the energy density of matter in the universe, E_m/V, is proportional to the cube of the photon temperature: $E_m/V \sim T^3$.

If we go back in time the universe contracts and the temperature of the photon gas increases. Eventually the photon energy, density which is proportional to the fourth power of the temperature, must overtake the matter energy density, which is proportional to only the third power. The transition to a radiation-dominated universe occurs when the universe is one-thousandth the size it is today and when the temperature is about 3,000 K, instead of about 3 K today. That temperature is beyond the melting point of most metals—a hot universe indeed.

Although matter dominates radiation today, we find a different story if we compare the entropy in matter with that in radiation. For a gas in equilibrium the total entropy is proportional to the total number of particles. Let us compare the entropy of matter (essentially the total number of nuclear particles out of which the galaxies are made) with the entropy of the photons (which is proportional to the total number of photons). The number density of nuclear particles—protons and neutrons—in the universe today is about 1 nuclear particle per cubic meter. (There is considerable uncertainty in this number—it could be ten times as large—but that uncertainty will not affect our description significantly.) The number of photons per cubic meter is about 400 million, a number determined by the universe's current temperature of 3 K. Hence the ratio of the entropy in the photons to that in nuclear matter (which is independent of the volume)—what is called the *specific entropy*—is 400 million (with uncertainties of a factor of about 10). We see that the entropy of the universe today is almost all in the radiation gas of photons, not in the matter.

The value of specific entropy is extremely important because it determines the nature of the universe. If the specific entropy were hundred of times as great as it is, one could show that the early universe would have been too hot to form galaxies and thus stars would not exist today. On the other hand, if the specific entropy were much smaller than its value today, almost all the hydrogen would have been made into helium in the big bang. Stars could then exist, but stars made only of helium are not very luminous. We conclude that if the specific entropy were very different from its value today, then the universe would be extremely different and probably hostile to the evolution of life.

The universe is a closed system, and hence its entropy, which we see is mostly in the gas of photons, increases in time in accord with the second law of thermodynamics. Galaxies form and stars burn, thus dumping photons into space and adding to the preexisting gas of photons. These processes increase the total entropy of the universe. But the remarkable fact is that the increase in the total entropy of the universe, from all these processes integrated over the entire lifetime of all the galaxies and stars, is only one ten-thousandth of the entropy already in the background photons—a tiny fraction. For all intents and purposes, the entire entropy of the universe today is in the photon gas and has remained effectively constant since the big bang. Entropy is essentially a conserved quantity in our universe.

Not so long ago scientists spoke of the "heat death" of the universe. In the 1930s the physicist James Jeans, reflecting the views of most of his colleagues, remarked:

> For, independently of all astronomical considerations, the general physical principle known as the second law of thermodynamics predicts that there can be but one end to the universe—a "heat death" in which the total energy of the universe is uniformly distributed, and all the substance of the universe is at the same temperature. This temperature is so low as to make life impossible.

Physicists like Jeans who realized the universe was subject to the second law of thermodynamics were not wrong about the heat death. But they did not know in the 1930s about the existence of the photon gas at 3 K. We now know that the "heat death" of the universe *happened long ago*—with the big bang that created the photon gas. Almost the entire entropy of the universe is in that photon gas. All the stars burning out can contribute but a tiny fraction to the total entropy that is already here.

Although the "heat death" of the universe is not the problem it once was, our new knowledge of the universe creates different problems. Since total entropy in an adiabatic expansion is conserved, the entropy of the universe has always been enormous. Where did it come from? And why is the specific entropy of the nuclear particles so small compared with that in the photon radiation? If this specific entropy were very different, then the universe would also be very different today. These questions, although I raise them here for the first time, will not be answered now. We will return to them when we discuss a pre–big-bang state of the universe called the "inflationary universe." Many physicists think that this pre–big-bang state holds the answers to many such outstanding puzzles which are not resolved by the big-bang picture.

By examining these problems of thermodynamics and cosmology, we begin to see an emergent theme in our study of the cosmos: the close relation between the smallest things we know, the quantum particles filling the universe, and the dynamics of the entire universe, the largest thing we know. This theme reaches its full significance as we now turn to examining the origin of the universe.

5

The Big Bang

I cannot deny the feeling of unreality in writing about the first three minutes as if we really know what we are talking about.
—Steven Weinberg

Let us imagine that we have a supercomputer that has programmed into it all the laws of physics as we know them today. The program contains the standard model of quarks, leptons and gluons along with some input numbers obtained from experiments like the masses of the quarks and leptons and the interaction strengths of the gluons. Using these data, the supercomputer can calculate the properties of the hadronic particles, determine how they scatter from each other and then build a model of nuclei and atoms.

The supercomputer also has programmed into it the Einstein equations. It deduces that there are three homogeneous and isotropic spaces that could describe the whole universe—the FRW cosmologies. But we have to tell the supercomputer which of these three FRW cosmologies applies to our universe. For definiteness we will tell it that the cosmic parameter $\Omega = 1/10$ corresponds to the open FRW cosmology. If the total matter density is a bit smaller or larger than this value, our supercomputer informs us that the computations on the early universe are not dramatically altered.

Finally the supercomputer has programmed into it the laws of statistical mechanics and thermodynamics. We tell the computer that for very early times in the history of the universe it can treat the universe as a homogeneous gas of quantum particles governed by the laws of statistical mechanics. This is an immense simplification that results in a huge saving of computer time. The computer determines that such a gas is in approximate equilibrium and so to compute all it needs to know is the temperature of the gas, the specific entropy of the various quantum particles, the conservation laws for the interactions of the quantum particles and their masses. We give it these input data and it is ready to go.

In short, our supercomputer simulates the universe much in the way that computers used by astrophysicists simulated the evolution of stars. Like Galileo's telescope in an earlier age, computers, because of their capacity to manage complex information, open a new window on reality. They show us a picture of a world we would otherwise never see.

While the supercomputer is useful in giving us a quantitative model of the universe, it is also useful for us to have a simple visual picture in mind to help us interpret its output, especially for the big-bang period. The big bang should *not* be visualized as an explosion that originates at a point in space and expands outward. A better way of visualizing the big bang is to imagine that the space of the universe is closed and is just the two-dimensional surface of a sphere. On the surface of that sphere is the homogeneous gas of quantum particles at a definite temperature which interacts according to the laws of statistical mechanics. The expansion or contraction of the universe is visualized as the expanding or contracting of the sphere. As the sphere contracts in time the gas on its surface gets hotter, and if it expands it gets cooler. Of course, if one assumes the universe is open, then instead of the closed surface of a sphere one has to imagine an infinite surface. The main point, however, is that the big bang is spatially homogeneous and isotropic—it happens everywhere at once, all over the universe. There is no "outside" to the universe where we can comfortably sit and watch the universe go through its evolution.

Now we run the supercomputer and it calculates the properties of the universe as it evolves in time. We can examine its output, displayed on screens and graphs, at our leisure and see in detail what is going on. One thing we notice right away is that its most interesting output is for very early times measured in minutes, seconds and microseconds. That is because for those very early times the temperature of the gas of photons has risen sufficiently for it to interact significantly with matter. After those initial hot early times, for billions of years to the present day, not much goes on from the view-point of microscopic quantum-particle physics. During those latter times the all-important macroscopic structures—galaxies, stars, planets and life—are made out of the primordial gas.

Examining the very early universe, we learn that the basic parameter that governs the physical processes is the temperature of the gas of interacting quantum particles that fills the whole space of the universe. Temperature, because it is proportional to the average energy of the colliding particles, establishes which new quantum particles can be created from the energy of the collisions. For particles of a certain mass to be created out of pure energy, a minimum threshold energy is required. Such energy thresholds are observed in high-energy-accelerator experiments for which a minimum energy is needed in order to produce new particles. These specific energy or temperature thresholds can be computed from the known mass-energy of the quantum particles observed in the laboratory. Since the temperature of the universe increases as we go backward in time, the existence of these temperature thresholds for particle creation implies that the early universe can be viewed as a series of stages or eras, each separated from the last by such a threshold.

For example, consider the threshold that occurs at the beginning of the "lepton era," when the universe is only about one second old and the temperature about 10^{10} K.★ Below this temperature the universe consists of mostly a radiant gas of photons. But as the universe heats above this temperature, something new happens. The colliding photons become so energetic that pairs of photons collide and convert themselves into massive electron-positron (antielectron) pairs. We know precisely the temperature at which this process first takes place because we know that the minimum energy of a photon that accomplishes this transformation (which is proportional to the photon temperature) is just equal to the mass of the electron, a known quantity, times the speed of light squared—the Einstein $E = mc^2$ equation is used. Of course, the electrons and positrons will annihilate back into photons almost as soon as they are made, but they hang around long enough to influence the dynamics of the gas.

This picture of particle-antiparticle production will be a basic theme of the big-bang story as the temperature further increases, beyond the threshold of the lepton era at 10^{10} K. At even higher

★ Here, and in what follows, all the temperatures and times will be approximate.

A History of the Universe in 9 Eras

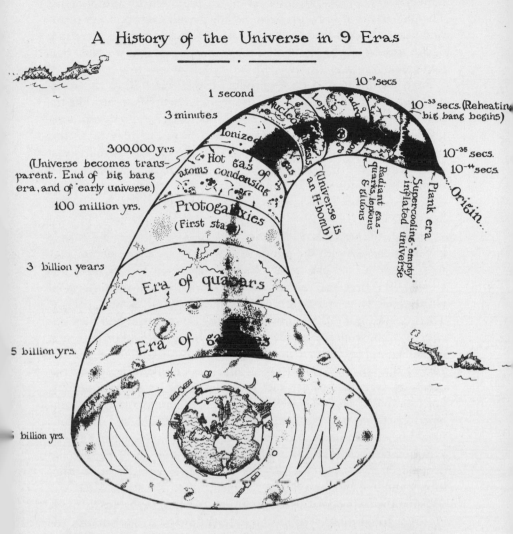

3 minutes

1 second

10^{-9} secs

10^{-33} secs. (Reheating big bang begins)

300,000 yrs.
(Universe becomes transparent. End of big bang era, and of early universe.)

10^{-35} secs.
10^{-44} secs

100 million yrs.

Ionized

Nucleosynthesis

Leptons

Hadrons

Hot gas of atoms condensing

(Universe is an H-bomb)

Radiant gas—quarks, leptons & gluons

Supercooling "empty" inflated universe

Planck era

Origin

Protogalaxies
(First stars)

3 billion years

Era of quasars

5 billion yrs.

Era of galaxies

5 billion yrs.

NOW

The history of the universe can be seen as a succession of eras. Shortly after the very origin of the universe (presumably out of nothing) it undergoes an inflationary period—its size expands enormously and it supercools. This is followed by the reheating phase, the creation of the gas of quantum particles and the beginning of the big bang—a series of eras during which hadrons and atomic nuclei get made. The big bang lasts until the universe is about 300,000 years old. Then recombinations of electrons with nuclei form the first atoms and the universe becomes transparent. Subsequently, stars and galaxies are formed.

temperatures, muon-antimuon pairs are produced by the photons. The universe as it heats up becomes filled with every kind of quantum particle and its antiparticle—a scene of vast carnage and creation. There are several important features of this picture that we must bear in mind.

First, the quantum particles produced, while each has a characteristic rest mass, may be treated as if they were massless—just like the photons—once the temperature of the universe is significantly greater than that rest mass-energy. The reason one may make this useful approximation is that the particles are moving so rapidly at high temperature that almost all their energy is in their kinetic energy of motion and not their rest mass-energy. Effectively, material particles become like radiation—massless, and moving at the speed of light.

Second, the new particles, once created, share the total available big-bang energy with the photons. For example, once the temperature threshold for the production of electrons and positrons is crossed, the universe consists of approximately equal numbers of photons, electrons and positrons, each having about the same energy. This equipartition of the numbers of different particles and their energy is a consequence of the universe's being in equilibrium as it expands—the rate of particle collisions is greater than the expansion rate of the universe. Then the available energy can be spread out evenly to each class of particles that participates in the interactions. For example, if we imagine that for a split second photons outnumbered electrons and positrons, then more electrons and positrons would be created until a balance, or equilibrium, was achieved. Steven Weinberg compares this balancing act to the law of supply and demand in classical economics—in a world of perfect competition, supply and demand always achieve equilibrium, the stable situation.

Using the laws of statistical mechanics (with the appropriate modifications to take into account the different quantum statistics for the integer- and half-integer-spin particles) one can determine precisely the number of particles per unit volume for each of the various quantum particles in equilibrium at any time during the big bang. The power of statistical mechanics is that we can determine such numbers solely from the fact that the particles are in equilibrium—the details of the complicated interactions need not concern us.

The importance of the exact-conservation laws, like the charge conservation, lepton-number conservation and baryon-number conservation laws which I previously discussed, also becomes apparent. Imagine all the quantum particles interacting at some very high temperature, and then the temperature falls as the universe expands. As it falls, we cross a particle-production threshold, and then those particles may cease to exist altogether. For example, the electrons and positrons almost all annihilate into photons when the temperature falls below their production threshold. But because today there is a very small excess of electrons, it must have survived the final electron-positron carnage as the temperature fell below the electron-positron threshold. This means it must have been there all along because of the rigorous law of charge conservation. Electrons have minus one unit of electric charge, while positrons have plus one unit. Only if there are more electrons to begin with is there an excess of negative charge. Then electric-charge conservation guarantees that some electrons must and do survive. Much, much later in the big bang, these excess electrons will combine with nuclei to form the first atoms.

We have seen that because of the existence of threshold energies, the big bang neatly organizes itself into a series of eras, each separated from the last by such a threshold. What happens during those eras depends crucially on the temperature range that characterizes the era and on the specific model of the quantum particles we have programmed in our supercomputer. The model we first examine is the "standard model." Let us start our supercomputer at the beginning of time and run forward.

THE SINGULARITY

According to the standard model, at time zero (by definition) the universe had infinite matter density, infinite curvature and infinite temperature—a state known as "singularity." A singularity sounds rather monstrous, even mysterious, and physicists have asked, Is such a singularity avoidable?

To answer this question, the mathematical physicists Roger Penrose, Stephen Hawking and George Ellis showed that under very general conditions (for example, that the universe, considered as a gas of particles, always had positive mass density and pressure), every solution to the Einstein equations must eventually develop a singularity—a state for which the universe has collapsed to a mathematical point—a result known as the "singularity theorem." While their work did not prove that these extreme conditions were really present at the beginning of time, the standard model certainly satisfies the requirements of their "singularity theorem." This means that if one adopts the Einstein equations along with some general conditions on the matter in the universe, then a singularity is inevitable.

The appearance of such a singularity is a good reason for rejecting the standard model of the very origin of the universe altogether. But this does not mean that it fails to provide a good model for particle interactions well after the very origin, once the density of matter has a large but finite value.

These singularities appear unambiguously in the mathematics, but do they really occur in nature? Even classical physicists found such singularities in their mathematical descriptions of nature—for example, an electrically charged point particle has an infinite energy density in the electric field at the point. But on the basis of past experience, such singularities in the mathematical descriptions of physical entities simply reflect an incomplete physical understanding. The appearance of mathematical singularities in the description of nature is really a challenge to physicists to devise a better mathematical description based on deeper physical laws that avoid the singularity. The singularity at the origin of the universe implied by some models should be seen as a challenge, not a veil of ignorance behind which we may not look.

According to the standard model, after the initial singularity the density of matter and temperature of the universe are enormous but finite. Both continue to drop rapidly as the universe expands. The model implies that the radiant gas of interacting quantum particles consists of the interacting quarks, leptons and gluons, all with a huge energy that allows them to freely convert into one another in a manner consistent with the conservation laws. Colored gluons convert into quark-antiquark pairs, which almost immediately annihilate back into gluons. The weak gluons convert into lepton-antilepton pairs, and so on—a vast scene of creation and destruction of all the quanta the standard model.

It is actually a simple universe without structure, a totally chaotic gas and very uniform. Because of that simplicity it can be easily managed mathematically. Not much of interest happens in the standard model of the universe until the temperature drops to about 10^{15} K—still an enormous temperature, way beyond the temperature of the interior of a star. But 10^{15} K corresponds to a mass-energy equal to that of the W and Z weak bosons, the largest mass scale in the standard model and the first of several energy thresholds we will cross. At this temperature the universe is about a tenth of a nanosecond (one-tenth of one-billionth of a second!) old.

THE THRESHOLD OF ELECTRO-WEAK SYMMETRY BREAKING: 10^{15} K

At temperatures above 10^{15} K, the electromagnetic and weak gluons interact symmetrically. As the temperature falls below about 10^{15} K, the symmetry breaks and the distinction between these two interactions becomes manifest—the weak bosons, W and Z, fall out of equilibrium with the other particles in the quantum soup because they are too massive to be created, while the photons remain in the soup because they are massless and easily created.

The distinction between the electromagnetic and weak interactions according to the Weinberg-Salam model is in part a consequence of a spontaneously broken symmetry. As an example of such a broken symmetry, I have described the alignment of all the small magnetic domains in a magnet producing a net magnetic field—the Heisenberg ferromagnet. But if we heat an ordinary magnet, its magnetic domains become agitated and disoriented and start aligning in random directions. At a certain critical temperature, the whole magnet completely loses all trace of magnetism because the domains no longer align in a preferred direction: the original rotational symmetry for which there is no preferred direction has been restored. This example reveals an important property of spontaneously broken symmetries: at some critical temperature they become restored.

The spontaneously broken symmetry of the Weinberg-Salam theory is no exception, and like that of the magnet, its broken symmetry is restored at a critical temperature, as was first emphasized by the Soviet physicists D. A. Kirzhnits and Andrei Linde. But unlike that of the ferromagnet, this temperature is so high (10^{15} K) that it could have been achieved only before the first nanosecond of the big bang. Above that critical temperature, the distinction between electromagnetic and weak interactions is unimportant. The weak gluons W and Z become effectively massless quanta like the photons, colored gluons and other quanta. The transition to the symmetrical situation at the critical temperature is rather smooth. As in the case of the magnet, as the temperature increases one simply sees less and less of the broken symmetry until at the critical temperature it vanishes completely and the original symmetry is restored.

Here, for the first time, we see a remarkable feature of the modern theory of the origin of the universe: the further back in time we go, the hotter the universe becomes, and broken symmetries are restored. The universe and all its particle interactions are becoming more and more symmetrical as we descend deeper into the big bang. This feature holds out the hope that the universe becomes simpler, more symmetrical and more manageable in its very early history, a hope to which physicists cling in their model building.

Conversely, were we to progress forward in time, we would see that as the temperature falls, those perfect symmetries are broken. Now the physical differences between the various interactions—strong, weak and electromagnetic—become apparent.

The universe today, with its relatively low temperature, is the frozen remnant of the big bang. Like an ice crystal that has frozen out of a uniform water vapor, it has lots of structure—the galaxies, stars and life itself. But according to the modern view, even the protons and neutrons—the very substance of matter—are the frozen fossils of the big bang. They too were created as the temperature fell. This event is called "hadronization."

HADRONIZATION: 10^{14} K

After the electro-weak symmetry is broken, the universe consists of a gas of approximately equal numbers of leptons, quarks, their antiparticles, colored gluons and photons being continuously created and destroyed. Effectively, the quarks are freely flying about and interacting with other particles—a brief period of parole.

Recall that quantum chromodynamics (QCD), the relativistic quantum-field theory that describes the interactions of quarks and colored gluons, has the property of "asymptotic freedom." At high energy the strength of the colored-gluon coupling becomes weaker —the stickiness of the gluons decreases. High energy corresponds to high temperature, and at the high temperatures beyond 10^{14} K the coupling strength decreases so much that the strong interaction becomes weak. At those high temperatures the hadrons literally became unglued and the quarks were freed.

However, as the temperature dropped below 10^{14} K and the universe continued to expand, gluonic jails—the little bags we call hadrons—formed about the quarks, imprisoning them for all future time. This transformation of the quantum-particle gas from one of free quarks and colored gluons to one of bound quarks or hadrons is what is called hadronization. It marks the beginning of the hadron era.

THE HADRON ERA: 10^{14} K to 10^{12} K

At temperatures below 10^{14} K the quarks are trapped in the hadrons, and the universe is about one-hundredth of a microsecond old. All the hadrons are now part of the quantum soup, and the free quarks and colored gluons have completely disappeared. It was as if a color movie (the colored quarks and gluons) all of a sudden became black-and-white (the hadrons).

Hadrons are the quantum particles associated with the strong force that binds the atomic nucleus together. The first hadrons physicists discovered were the nucleons—the proton and the neutron. Then pions—spin-zero particles, much shorter-lived than the neutron and only about one-seventh as massive—were discovered, and soon lots of other similar hadrons were found. Today all these experimentally detected hadrons are viewed as permanently bound systems of quarks.

At the high temperatures of the hadron era, the photons and other particles in the gas of quantum particles are sufficiently energetic to produce hadron-antihadron pairs. These particles share the total available energy with all the other particles. Hence, even as new particles come on the scene the *total* number of particles, which is proportional to the entropy, remains the same. Since the various particles, including all the hadrons, share the available energy at a given temperature, this means that there are approximately equal numbers of each different kind of particle.

For example, when the temperature is sufficient to create pions at the onset of the hadron era, there are approximately as many pions as photons, electrons, positrons, muons and so on. When the temperature is sufficient to create nucleons (about 10^{13} K), we conclude that the number of nucleons and antinucleons was approximately the same as the number of each other particle; in particular, the number

FOUR FUNDAMENTAL INTERACTIONS

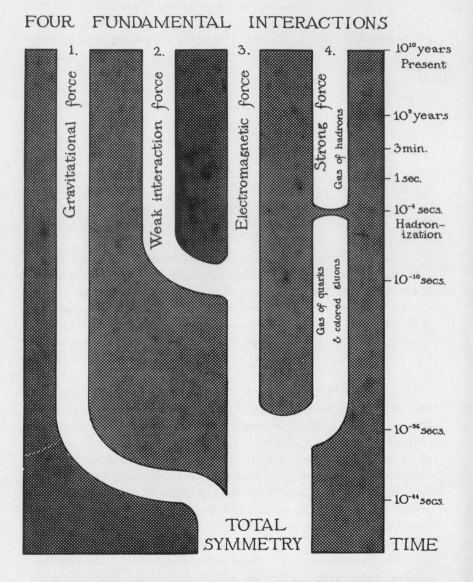

The evolution of the universe as a succession of broken symmetries. If the ideas of modern field theory are right and broken symmetries are restored at high temperature, then at the very earliest times the four known forces of nature, now seen as distinct, were unified.

of nucleons was about the same as the number of photons. This is really quite remarkable, if we recall that the photons today outnumber the nucleons by about 400 million to one. Therefore, at the end of the hadron era when the universe was about one ten-thousandth of a second old, all those "extra" nucleons annihilated with antinucleons, leaving but a tiny fraction of proton and neutron survivors that have lasted until the present day. Why did they survive?

The standard model has a strict baryon-number-conservation law which implies that in any particle interaction the number of baryons *minus* the number of antibaryons is conserved. Since the nucleons, the proton and neutron, are the lightest baryons, all other baryons can disintegrate into them, passing on their baryon-number charge. The neutron can further decay into a proton and pass its baryon number on to it. But this neutron–decay process takes about a thousand seconds, and that is long compared with the entire duration of the hadron era. That is why the neutron is effectively stable during this period.

The law of exact baryon–number conservation thus requires that if we end the hadron era with a tiny excess of baryons over antibaryons, that excess has to have been there all along, well before the hadron era. This tiny excess is today reflected in the large specific entropy of the universe—the excess of the number of photons over that of nucleons. The existence of nucleons, the visible matter of the stars and galaxies, seems like an accident, a lucky remnant from an earlier era of the universe.

Physicists, in their attempt to comprehend the universe, have tried to explain the tiny excess of nuclear matter over antimatter. They start with the assumption that the universe began in a symmetrical state in which the baryon number was effectively zero. But if the standard model is correct and baryon number is conserved, then the baryon number today would also be zero—a disaster because then there would be no visible matter in the universe. These physicists therefore appeal to GUTs (Grand Unified Theories) that go beyond the standard model and imply that baryon number is not conserved. One consequence of GUTs is that the proton can decay—and if it can decay, it can also be created. Another consequence was first pointed out by the Soviet physicist Andrei Sakharov, even before the GUTs were invented. It is that the tiny baryon excess can indeed be created from a universe with zero net baryon number, provided that the gas of interacting quantum particles had special properties. Such ideas take us well beyond the standard model I am describing here, but we will return to them in a future chapter.

As the temperature decreases during the hadron era, many energetic thresholds, corresponding to the masses of various hadrons, are crossed. Some hadrons fall out of equilibrium with the other particles and, unless a conservation forbids it, they are utterly annihilated. By the time the temperature has fallen to about 10^{12} K, most of the heavier hadrons have annihilated (with the exception of protons and neutrons) and we now stand at the threshold of the lepton era.

THE LEPTON ERA: 10^{12} K to 10^{10} K

At the start of the lepton era the universe is one ten-thousandth of a second old, the temperature is 1 trillion Kelvin (10^{12} K) and each cubic centimeter of the cosmic quantum soup weighs about a thousand tons. The universe consists of a mixture of approximately equal numbers of photons, electrons, electron neutrinos, muons, muon neutrinos, some other particles like pions—light-mass hadronic relics of the previous era—and their antiparticles, plus a relatively small "contamination" of equal numbers of protons and neutrons which are no longer in equilibrium with the other particles. But the small number of protons and neutrons do continue to interact with all these other particles. For example, a proton when it interacts with an electron is converted into a neutron plus an electron neutrino. Neutrons, when they interact with positrons, convert into protons and antielectron neutrinos. Although the total number of protons and neutrons is very small—one per roughly 100 million of the other particles—the protons and neutrons are rapidly converting into one another because of their interactions with the leptons.

As the temperature falls from its value at the beginning of the lepton era, the production threshold for the muons is crossed. All the muons and antimuons now annihilate into electrons, positrons and muon and electron neutrinos. Any excess charge of the muons can be passed on to the electrons (the electron, as you will recall, is the lightest charged particle and has no other lighter particle to pass its charge to). For this reason no muons survive the muon slaughter. But muon neutrinos, since they carry muon lepton number—a conserved quantity—even though they now cease to interact with the other particles must continue to exist. Furthermore, their number density is approximately equal to that of the photons, since that is what it was just before they stopped interacting. Vast numbers of muon neutrinos are now free to wander about the universe interacting hardly ever, just as the photons do today.

Likewise, at a still lower temperature the electron neutrinos fall out of equilibrium with the electrons and positrons. They then join the muon neutrinos along with tau neutrinos (which fell out of equilibrium even earlier than the muon neutrinos) to fly about the universe, a noninteracting gas. As the universe expands, the wavelength of these neutrinos gets red-shifted and, as with the photons, their temperature falls. Today the neutrinos are a background radiation gas just like the gas of photons, filling the universe at a temperature slightly lower than that of the photons, about 2 K. The reason for the slightly lower temperature is that the neutrinos stopped interacting before the electrons and positrons annihilated into photons. This annihilation process, when it takes place, heats up the photon gas by about 30 percent over that of the neutrinos.

If this picture is correct, then why don't scientists detect the neutrino gas? Unfortunately, these relic neutrinos, in spite of the fact that there are hundreds of them per cubic centimeter today, interact hardly at all. Their interactions are so extremely weak that it would take improvement by a factor not of 10 but of millions on current neutrino-detecting technology to find them. Yet physicists are studying this problem, and perhaps someday a clever method of detecting these relic neutrinos may be found. It is clear, if these ideas are right, that the bulk of the present entropy of the universe resides not only in the photon gas but also in the neutrino gas.

These free neutrinos are usually thought of as strictly massless. An intriguing possibility is that they may in fact possess a small mass. If that is the case, then the relic gas of neutrinos could be the dark matter that might close the universe. Experimental physicists have tried to measure the electron neutrino's mass and concluded that it cannot be large enough to provide the dark matter. But the experimental limits on the muon and tau neutrino masses are much less restrictive, and these neutrinos could do the job.

If there are massive neutrinos, each with a mass about one five-thousandth of the electron's mass (it cannot be greater, because then the density of the neutrino gas would exceed the observed limit on the average mass density of the universe), then one can show that the mutual gravitational attraction among them will tend to form gigantic clusters with about the mass of today's superclusters of galaxies. A picture of supercluster formation then suggests itself. The massive neutrinos, liberated during the lepton era, begin to gravitationally cluster, growing into neutrino clumps the size of superclusters of galaxies. Later, after the big bang is over, hydrogen and helium gas will gravitationally fall into these gigantic neutrino clumps, forming pancake-shaped objects. These "pancakes" the size of superclusters of galaxies, through complicated interactions involving the hydrogen gas, then fragment into galaxy-sized objects. This happens in the first million to 10 million years after the big bang, the era of galaxy formation. I have discussed some of the consequences of this pancake picture of supercluster and galaxy formation in a previous chapter.

Whether or not this pancake picture is correct, we see here a fine example of the speculative interplay between particle physics and astronomy. The largest structures—superclusters of galaxies—may be telling us about some of the smallest structures—neutrinos and the properties of the universe before it was one second old. The whole universe becomes the proving ground for fundamental quantum physics.

The lepton era, besides liberating the neutrinos forever, also establishes a neutron–proton ratio of 2 neutrons to every 10 protons, an important ratio for establishing the final amount of helium that gets made during the next era—the photon era. At the beginning of the lepton era there are equal numbers of protons and neutrons because they are freely converting one into the other. But a neutron is slightly more massive than a proton—about 0.14 percent—and it can decay into a proton, an electron and an antielectron neutrino. By the end of the lepton era the temperature has fallen enough so that the small mass difference between the proton and neutron makes an important difference in their relative numbers. Because of this small mass difference it becomes more likely for a neutron to convert into a proton than vice versa. After detailed calculations, physicists conclude that there are only 2 neutrons for every 10 protons at the end of the lepton era when the temperature has dropped to 10 billion Kelvin (10^{10} K).

Some theoretical physicists who have done these detailed calculations emphasize that this neutron–proton ratio depends crucially upon the number of different kinds of neutrinos. And since the amount of helium that is made depends directly on this ratio, it also depends upon the number of different neutrinos. According to the calculations, if more than four neutrinos existed, then too much helium—about a few percent above the observed amount—would have been produced. Right now, as in the standard model, there are only three different neutrinos—the electron, the muon and the tau neutrino—so the calculations assuming the standard model give the observed amount of helium. Other physicists think that the calculational uncertainties and the uncertainty in the estimated amount of primordial helium observed in the universe today suggest that a rigorous conclusion severely limiting the number of neutrino species is unwarranted. Nonetheless, these estimates again underscore the intimate relation between the properties of the observed universe (the amount of helium) and fundamental quantum physics (the number of neutrino species).

THE PHOTON ERA: 10^{10} K to 10^3 K

At the end of the lepton era all the heavy leptons, muons and tauons have disappeared, while hordes of neutrinos flood the universe but no longer interact with anything. The only hadrons remaining are a tiny contamination of protons and neutrons with 10 protons for every 2 neutrons. Photons, electrons and antielectrons are still in equilibrium, creating and destroying one another. When the temperature falls below the production threshold to create electron-positron pairs, most of these pairs annihilate into photons (reheating the universe slightly). This temperature threshold marks the beginning of the photon era. Positrons are removed from the soup by annihilation, and the small number of negatively charged electrons remaining is equal to the number of positively charged protons (assuming the total electric charge of the universe, a conserved quantity, was zero initially). Since there is only about 1 proton per 400 million photons, it follows that there is one 1 electron per 400 million photons. The universe is now radiation, dominated by the photons—which continue to interact—and the neutrinos, which do not interact.

At the first second (which marks the beginning of the photon era, which goes on to last for 300,000 years), the temperature of the photons was 10 billion Kelvin and the density of the radiation about 100 kilograms (about 220 pounds) per cubic centimeter—a very thick, viscous fluid of light. The entire universe is about to become a gigantic thermonuclear reactor. In the next 100 seconds or so, almost all the helium we observe today gets made, by fusion burning of hydrogen (protons).

The stars, which formed well after the big bang, also burn hydrogen into helium, but they do so at a much slower rate. Since 10 billion years ago, when the first stars came into existence, only 2 or 3 percent of the hydrogen in the universe has been converted into helium by burning stars. Yet 25 percent of all the visible matter in the universe consists of helium which was made in a few minutes during the big bang. The hydrogen fusion producing helium released energy like a hydrogen bomb. But even the contribution of that immense energy to the total energy density already in the photon gas was minuscule and did not significantly reheat the universe.

At the beginning of the photon era there is a small contamination of protons, neutrons and electrons in the thick fluid of pure light— what remains of the quantum-particle soup. The number of protons is equal to the number of electrons, so that the total electric charge is zero and there are still about 2 neutrons for every 10 protons.

During the first few seconds, the protons and neutrons are continually bombarding each other and the photons. As protons and neutrons collide they can form the deuterium nucleus, consisting of a single proton and a single neutron bound together. Deuterium has a very loosely bound nucleus—the proton and neutron are easily liberated when struck by the ever-present photons. During the first few seconds, deuterium nuclei are torn apart as rapidly as they get made.

In contrast to deuterium, the nucleus of helium, consisting of 2 protons and 2 neutrons all stuck together, is tightly bound—it takes a lot of energy to tear a helium nucleus apart. Once it gets made, it is here to stay for quite a while. One can easily make a helium nucleus from two colliding deuterons sticking together. The problem with making helium by fusing deuterium during the first few seconds is that deuterium is so much less stable—it gets unmade at the same rate at which it gets made, and so there isn't much deuterium around. This is the "deuterium bottleneck" to the production of helium.

After about 100 seconds, the temperature has dropped to 1 billion K. Suddenly the photons are insufficiently energetic to break up the deuterons when they form. But a neutron, if free and not bound up into nuclei, as is mostly the case here, decays into a proton, an electron and an antielectron neutrino in about 1,000 seconds—a time only a factor of 10 larger than the age of the universe at this point. So some of the neutrons (initially 2 for every 10 protons) have now had a chance to decay into protons by the time the universe is 100 seconds old. As a result, out of every 16 nuclear particles 14 are now protons and 2 are nuetrons. The 2 neutrons can form two deuterium nuclei by joining with 2 protons. Now that the temperature of the universe has dropped, the deuterium bottleneck is gone and deuterium is sufficiently stable for it to collide and form helium. Almost all the deuterium fuses rapidly into helium, and by the time the universe is about 200 seconds old the fusion burning process is complete. Out of the original 16 nuclear particles, 4 of them, 2 neutrons and 2 protons, are bound up into helium, while the remaining 12 are protons. We see that 4 out of 16 nuclear particles, or 25 percent of the nuclear matter in the universe, is helium, and most of the rest is hydrogen. This is precisely what is observed today—a quantitative result of the first few minutes of the universe, and a powerful confirmation of the big-bang idea.

We see that the amount of helium produced depends primarily upon the initial proton–neutron ratio at the beginning of the photon era and also on the rate at which the temperature falls. This amount is not especially sensitive to the ratio of the number of photons to the number of nuclear particles—the specific entropy of the universe.

However, a small amount of deuterium, only about one-hundredth of 1 percent of all the hydrogen (a ratio which is observed today), evidently escapes. It does not get fused into helium. Interestingly, this small, trace amount of deuterium does depend rather sensitively upon the specific entropy—the ratio of the number of photons to nuclear particles. When the number of nuclear particles is relatively high (low specific entropy) there are more deuterons to collide with each other, and few survive the thermonuclear holocaust. But if the number of nuclear particles is relatively low (high specific entropy), then fewer deuterons are around to collide and make helium, and more survive. By keeping "a low profile" more deuterium escapes burning.

The fact that the observed relative deuterium abundance is high (one-hundredth of 1 percent is *high*) implies that the specific entropy is high—about 400 million photon particles per nuclear particle. Some of the deuterium produced in the big bang can be destroyed by falling into stars in the subsequent evolution of the universe. So it is even possible that more deuterium may have been produced in the big bang than we observe today.

Because of such uncertainties, most physicists and astrophysicists think that the amount of deuterium observed today is a lower limit on the amount produced during the photon era. But if that is so, then we conclude that the value of the specific entropy implies a current density of nuclear, visible matter corresponding to a cosmic parameter $\Omega = \frac{1}{10}$—not enough to close the universe. If we want a larger value for Ω, then there must be dark matter, possibly massive neutrinos or other exotic particles outside the framework of the standard model. Here we have yet another example of how the physics of the microworld—the deuterium production in the photon era—has cosmological implications: the value of the cosmic parameter Ω.

After the first few minutes, the immense thermonuclear reactor that is the universe shuts off. Nucleosynthesis is now completed; the temperature continues to fall as the universe expands. The universe now consists of a gas of photons, electrons, protons and the nuclei of the light elements like helium and deuterium. Not much happens in this plasma state (similar to the interior of a star) until about 300,000 years have elapsed and the temperature has fallen to a mere 3,000 Kelvin. Then something spectacular happens: the universe becomes transparent. This event is called "recombination."

RECOMBINATION AND THE END OF THE BIG BANG

The first 300,000 years of the universe was a burning world of darkness; it was opaque to the transmission of light. The universe was similar to the interior of the sun, which is also opaque—you cannot see directly through the sun. If any electrons combined with protons or helium nuclei to make atomic hydrogen or helium, they would immediately be knocked out by the energetic photons. Hence photons don't get to travel very far before they interact. For this reason optical telescopes will never see light from events earlier than about 300,000 years, any more than they can see inside the sun.

But once the temperature falls below about 3,000 K, the electrons combine with the nuclei to form true atoms (this is the "recombination" event) because the photons are no longer energetic enough to knock them apart. Now the photons effectively cease interacting and are free to fly about at the speed of light in all directions. All at once the universe becomes transparent, bathed in a brilliant yellow light, the color corresponding to matter heated to 3,000 K. By convention, this event marks the end of the big bang and the structureless expansion of the universe; soon structures—the protogalaxies—will begin to emerge.

At almost the same time, another event of importance occurs: the energy density of matter in the form of atomic hydrogen and helium overtakes the energy density in the photons. The universe becomes matter-dominated, not radiation-dominated—a feature it retains to the present day, when there is a thousand times as much matter in the universe as radiation density.

After recombination the temperature of the universe continues dropping, and its color changes from yellow, to orange, to red, to a deep red, and then to the darkness of deep space. At an age of 10 million years, our computer tells us, the matter density was a million times what it is today, about one hydrogen atom per cubic centimeter. Then, in fact, the density of matter for the whole universe was equivalent to the density of matter in just the galaxies today. This implies that galaxies resembling those of today could not have existed when the universe was only 10 million years old—they would have been lying one right on top of another then. Galaxies or protogalaxies probably formed somewhere between the first 100 million years and 1 billion years as hydrogen and helium atoms fell into the lumps of preexisting invisible dark matter.

As we run our computer forward in time starting from the first million years—when the matter in the universe was a uniform gas of hydrogen and helium—to the first billion years, we can see the galaxies form. Huge lumps of hydrogen and helium gas would form out of the uniform gas. They would be either the size of superclusters (according to the "pancake model") or just the size of individual galaxies themselves (according to other models which I have discussed in a previous chapter). Population III-type stars—stars made of pure hydrogen and helium without heavy elements—might get made. Very massive stars would burn out quickly, collapsing into black holes or neutron stars; this would generate shock waves in the remaining gas, compressing it and thus creating the conditions for more star formation. Gigantic black holes might form in the nuclei of the galaxies, consuming stars and emitting enormous amounts of light—the first quasars. The universe is now well on its way to developing more and more complex structures—galaxies, stars, planets and eventually life: the inhabitants of Herschel's garden. The temperature of the photon gas continues dropping until it reaches the value of 2.7 K of today. While this low temperature is but an insignificant reminder of its former glory, the detection of this temperature was the breakthrough that lent credibility to the entire big-bang model I have just described.

BEYOND THE ELECTRO-WEAK SYNTHESIS

Looking over the whole lifetime of the universe, we see that the period cosmologists understand the best—the true big bang—lasts from about the first nanosecond—the electro-weak symmetry breaking—to the first 300,000 years—the time of recombination. Both before and after that big-bang period, things are not well understood. For example, the period of galaxy formation is hard to study because of its sheer complexity. Only the future deployment of new telescopes will provide the data scientists need to penetrate this complicated era. Likewise, the temperatures and energies before the breaking of the electro-weak symmetry are so high that they have not yet been duplicated in any high-energy physics labs. What goes on in that very early period is a guessing game for the field theorists.

Suppose we go back in time to the first nanosecond, and using our supercomputer, let time run backward so that the temperature increases. What happens?

According to the standard model, not much. The radiant gas of quarks, leptons and gluons continues to contract and its temperature increases. Since the density and pressure of this gas obey the conditions of the Penrose-Hawking singularity theorem, the singularity at the very origin of the universe is eventually encountered and our computer spews forth infinite numbers—just nonsense. In order to develop a picture of the universe before the first nanosecond, we need to go beyond the standard model of the quantum particles to a new model. How can we build a new model? What criteria must it satisfy?

The standard model of quarks, leptons and gluons has the virtue of being partly tested in high-energy laboratories. If we are to go beyond it into the even-higher-energy regimes encountered before the first nanosecond, we must then leave the secure ground tested and examined in our labs and venture forth into the unknown with our imagination to guide us. But not only our imagination. We can also approach the question logically. Important events must have happened before the first nanosecond to set up the proper conditions for the universe to evolve into what we see today. If we are not careful, the flights of our imagination will be quickly grounded.

At first one might think that building a new model that includes the standard model but also goes beyond it would be rather easy. But it is not. The difficulty is that if one is not very careful, the new model will predict a state of the currently observed universe which is at complete variance with the facts.

The present state of our universe depends critically upon certain physical quantities lying within a delicate range of values. I have already mentioned one such physical quantity, the specific entropy of 400 million photons per one nuclear particle. If that quantity was very different from its present value, then the universe as we observe it would not exist. In the standard model of the early universe, the value of the specific entropy is an input—it corresponds to the initial amount of baryon-number charge in the universe. Other models that might go beyond the standard model could determine the specific entropy but unfortunately yield the wrong value, thus leading to a universe that does not exist. Ambitious model builders must be careful.

Other examples of such critical physical quantities are the values of the quark masses. For example, the down quark has a heavier mass than the up quark, and for this reason the neutron, which contains more down quarks than the proton, is heavier than the proton. This implies that a free neutron may decay into a proton, thus releasing energy. But if instead the up quark were heavier than the down quark, the neutron rather than the proton would be the stable nucleon. But then the hydrogen atom could not exist, because its nucleus is a single proton and that could now decay into a neutron. About 75 percent of the visible universe is hydrogen, and we would conclude that it would not exist if the value of the quark masses were just slightly different.

There are many examples of such physical quantities which cannot lie outside a narrow range of values or the universe would not be as it is; stars, galaxies and life might not exist. From the viewpoint of the standard model these quantities are simply assumed to have their observed values. They are inputs into our supercomputer and, logically, they could have other values. But physicists want to understand the value of these specific constants as observed on the basis of a master physical theory and not just have them be inputs. Such a master theory, if it exists, clearly goes beyond the standard model, since it should logically fix those constants exactly. A master theory would fulfill Einstein's dream that "there are no *arbitrary* constants."

To realize that dream, ambitious theoretical physicists are exploring "wild ideas," ideas which currently have no experimental support but which aren't inconsistent with experiments either. These wild ideas at the frontier of current research may tell us about the time before the first nanosecond of the universe and perhaps reveal the very act of creation itself. The whole universe is their proving ground. Casting conservative caution aside, let us now look at these wild ideas.

Three

Wild Ideas

What is now proved was once only imagin'd.
— William Blake

1

Unified-Field Theories

It is a wonderful feeling to recognize the unifying features of a complex of phenomena which present themselves as quite unconnected to the direct experience of the senses.

—Albert Einstein, 1901

My first contact with unified-field theory came when I was thirteen. I read a newspaper article in 1953 which announced with great fanfare that Einstein had finally devised a unified-field theory, the culmination of his life's work. The story, titled "Einstein Offers New Theory to Unify Law of the Cosmos," even contained Einstein's unified-field equations, which were completely unintelligible to me. Yet the article communicated to me the idea that beyond the complexity of the world and the plurality of sensations there was a unifying order and that Einstein had somehow grasped it. This was simply wonderful—the essence of the universe, the mystery of existence was now embodied in a few equations. I became excited. Years later I learned that not much came of Einstein's unified-field theory and that most physicists think it is wrong. Einstein himself was embarrassed by the attention the article generated.

But my youthful exuberance generated something that I cannot let go of even today—the idea that a single, simple physical law accounts for the totality of material existence. Such a physical law would explain the origin of the universe, its contents and its destiny. All other natural laws could be logically deduced from this one law. If such a law were discovered, it would be the final triumph of physics: the logical account of the foundation of existence would then be complete.

No one, physicists included, has the slightest proof that such a master law exists. It is easy to imagine lots of problems. Perhaps the very idea of physical law breaks down at some point. For example, the mathematical description of nature, which so far has not failed physicists, is conceivably inadequate to the task of expressing such a law. Another possibility is that the master law exists but the human mind is incapable of finding it. Even an artificial superintelligence with capabilities beyond the human mind would be limited by the master law itself. Therefore it could not discover the master law—a kind of "Catch-22" in understanding the universe.

Physical laws can be compared to the rules of sports. But unlike sports rules, which are composed by human beings, physical laws seem to be inherent in the order of the universe, which people did not invent. Sometimes sports rules are changed by the players, who, for example, might allow for a handicap if there is a disparity in skill. In this instance there is an unwritten rule that governs the change in the rules—the rule that players want to make the game more challenging and interesting by evening the competition. Likewise one can imagine that the physical laws change, but then there is a new law that governs the change. Conceivably, as physicists discover new laws that logically subsume the previous laws, they may find that the process never terminates. Instead of finding an absolute universal law at the bottom of existence, they may find an endless regress of laws, or even worse, total confusion and lawlessness—an outlaw universe.

So there is no guarantee that a simple physical law awaits us. Yet in spite of this possibility, the notion of a simple law describing all existence beckons like the Holy Grail. And, like the hunt for the Holy Grail, the search may be more interesting than the object sought. How do physicists search for the laws of nature?

A FREE INVENTION OF THE MIND

Not so long ago, many reflective people thought that physicists logically deduced the laws of nature directly from experiments and observation. The basic laws were intimately related to experiment. Today this method has been abandoned and physicists do not directly deduce the laws from experiment. Instead they try to intuit the basic laws from mathematical reasoning. No one else has stated this shift away from strict empiricism as well as Einstein in his Herbert Spencer lecture in 1933. He remarked:

> It is my conviction that pure mathematical construction enables us to discover the concepts and the laws connecting them, which gives us the key to the understanding of Nature. . . . In a certain sense, therefore, I hold it true that pure thought can grasp reality, as the ancients dreamed.

Einstein was profoundly influenced by his own invention of the general theory of relativity. He created a purely mathematical construction, what we would call a model, a "free invention" of his mind, to describe the physical world. From this model he logically deduced several quantitative implications for experiment and observation—a small shift in the orbit of the planet Mercury, the bending of light around the limb of the sun and the fact that clocks should run more slowly in a gravity field. If observations fail to confirm these implications of the model, the model fails; hence it is a testable model. But the model itself is freely created and not deduced from experiment. Einstein went on to comment:

> If the basis of theoretical physics cannot be an inference from experience, but must be a free invention, have we any right to hope that we shall find the correct way? Still more—does this correct approach exist at all, save in our imaginations? To this I answer with complete assurance, that in my opinion there is the correct path. Moreover, that it is in our power to find it.

Finding that "correct path" is the ambition of the contemporary builders of field-theory models. That path seems to be leading them to the very beginning of the universe; whether it is a false path time will tell. In their recent attempts to comprehend the universe, theoretical physicists have "gone for broke." They are extending their theoretical models out far beyond the energies actually probed in the laboratory, to reach to the enormously higher energies encountered before the first nanosecond of the universe.

Most of these models, free inventions of their minds, are created by young scientists, with their concentrative vision, their boundless free energy and their remarkable ability to sublimate their most primitive impulses into the intellectual ambition to know. Physicists, in their conceptual card game with nature, have already won a few tricks and now want to "shoot the moon"—go all the way to the beginning of time. It's hard to tell whether they are bluffing or really have all the needed cards. A profound revision in our concept of material reality may be required before an explanation of the origin of the universe is possible. But it is already clear that relativistic quantum-field theories and their intricate symmetries are providing conceptual surprises, an unanticipated richness in explanatory power which has physicists excited. The theme of their work has been the unification of quantum fields and their corresponding forces through the application of symmetry principles.

At first it seems futile to try to reduce the diversity of the forces of nature to a single underlying simple force. Yet physicists have made considerable progress on this problem in spite of the obstacles. An early example of such a force-field unification was Maxwell's mathematical unification of electric and magnetic fields into a single electromagnetic field. Prior to Maxwell's work, electric and magnetic fields were seen as interrelated but distinct. After Maxwell, physicists realized that this interrelatedness was more profound than previously thought—electric and magnetic fields actually transform into each other as they change in time. If electric and magnetic fields oscillated in time they could propagate as an electromagnetic wave in space, a wave that could be identified with light. Maxwell's unification of electric and magnetic fields thus led directly to the remarkable discovery that light is an electromagnetic wave, answering the ancient question "What is light?" in a new way.

Field unification was but one aspect of Maxwell's discovery. Yet another was "parameter reduction." In their earlier studies of electricity and magnetism, experimental physicists had determined two physical constants—the "electric and magnetic susceptibility" of empty space. These two constants appeared in the electromagnetic-wave equations, and therefore Maxwell could calculate the velocity of the wave—the speed of light—in terms of them. Hence three different experimental parameters previously thought to be independent—the electric and magnetic susceptibility and the speed of light—were now related in a fixed and determined way. Instead of three independent parameters there were now only two. Such parameter reduction is another goal of the field-unification program. Ultimately the goal is to find a master theory with no arbitrary parameters, so that every physical constant can be calculated.

Physicists are far from realizing this ultimate goal, but the field-unification program moves steadily forward. Of the four observed forces—the gravitational, electromagnetic, weak and strong forces—physicists have devised theories which unify three, with gravity the odd man out. Work is in progress devising unified-field theories which also include gravity, although they are not yet very realistic.

GRAND UNIFIED THEORIES: GUTs

The modern initiative to find unified-field theories begins with Einstein's work in the 1920s and 1930s. Einstein, starting with both his theory of general relativity as describing gravity and Maxwell's theory as describing electromagnetism, sought a more embracing unified theory that would integrate both forces. At the time he did this work the strong and weak forces were only beginning to be understood, forces which today's physicists consider as fundamental as gravity and electromagnetism. Einstein's vision of the unified-field theory was that it would emerge as a consequence of the combining of quantum mechanics with general relativity.

Though Einstein's efforts to construct a unified-field theory failed, he inspired other physicists by showing them the possibility that all the diverse forces of nature might be manifestations of a single unified field. During the following decades when physicists were exploring the weak, strong and electromagnetic forces, the idea of unification was always in the back of their minds, beckoning like a promised land.

A number of physicists, including Julian Schwinger, Murray Gell-Mann, Sheldon Glashow, Abdus Salam, John Ward and Steven Weinberg, began emphasizing in the 1950s and '60s that the prospect of uniting the electromagnetic and weak interactions was more plausible than uniting the electromagnetic and gravitational interactions. This was because electromagnetic and weak interactions had an important property in common: they were both thought to be mediated by spin-one gluons, while gravity was mediated by a spin-two gluon. These intuitions bore fruit when the electro-weak model was invented in 1967–1968, and the field-unification program took a profound new direction. The royal road to unification was now seen as uniting fields under the aegis of a spontaneously broken symmetry. The notion of a broken symmetry explained how it was possible that forces which were fundamentally unified and symmetrical could be manifested so differently in nature. The electro-weak model when taken together (although not unified) with quantum chromodynamics, the theory of the strong force, make up what became known as "the standard model" of interacting quarks, leptons and gluons.

During the 1970s, as experimental evidence accumulated in favor of the unified electro-weak model, some physicists, taking the eventual experimental success of the model for granted, were already trying to mathematically extend the unification scheme to include the strong quark-binding force mediated by the colored gluons. They wanted a synthesis of the weak, electromagnetic and strong forces. Since the colored gluons, like the electromagnetic and weak gluons, are all quanta of Yang-Mills gauge fields, it is natural to imagine that all gluons in the standard model are but components of a single, unified field. This is the basic modern idea of field unification.

Field theories that indeed unify the three forces—the electromagnetic, weak and strong forces—are called "Grand Unified Theories," or GUTs. By intentional design, GUTs fall short of total field unification because they do not include gravity, by far the weakest of the four known forces.

Theoretical physicists exploring GUTs today believe they have profound implications for the nature of the very early universe even before it was a nanosecond old. GUTs also imply the existence of new properties of quantum particles such as proton decay (protons are stable in the standard model) as well as the existence of a whole new class of quantum particles, the magnetic monopoles.

Yet as exciting as GUTs are, unlike their progenitor the electro-weak model they have no direct experimental support. GUTs are examples of wild ideas—new ideas that are consistent with general physical principles and existing experiments but as yet have no direct supporting evidence.

Of all the wild ideas I will describe, that of GUTs is certainly the most mature. Theoretical physicists have concocted a number of ingenious GUTs, specific mathematical models, to calculate the properties of proton decay, and their experimental colleagues are trying hard to find evidence for this decay process. So far it hasn't been seen. If the GUT idea turns out to be wrong—as well it might—it will be very disappointing and turn physicists toward a new direction.

The common feature of the strong color force and the electro-weak interactions is that they are both mediated by gluons—the quanta of Yang-Mills fields, fields which are a consequence of symmetry. The problem of unifying these interactions is thus the problem of finding a single (rather than a multiple) symmetry that upon spontaneous symmetry breaking results in the smaller subsymmetries that correspond to the strong and electro-weak interactions.

A way of envisioning this process of unification and symmetry breaking is to symbolize the symmetry of the colored strong force by a circle, which is symmetrical about its center. Likewise the symmetry of the electro-weak model is represented by a separate circle. We could imagine that the radii of these two distinct circles are inversely proportional to the strength of the corresponding interactions. The two circles have nothing directly to do with each other, just as the symmetry of the strong interaction and the symmetry of the electro-weak interaction are completely independent of each other in the standard model.

The idea of field unification can now be grasped if we imagine that these two apparently independent circles are actually different great circles—equators—on a single sphere. A circle is symmetrical only about a point, while a sphere, symmetrical about any axis, has more symmetry. The previous circles are now seen as just subsymmetries of the spherical symmetry, which incorporates both and unifies them. Since they both have the same radius, the strong and electro-weak interactions have the same strength in this picture—they are unified.

Spontaneous symmetry breaking can also be understood in the context of this image. The perfect spherical symmetry, although a solution to the field equation, is not stable: the sphere squashes into a ellipsoidal figure, which is the stable solution. The squashed sphere may be characterized by two circles of different radii corresponding to the different strengths of the various interactions—a manifestation of the broken symmetry. But the underlying starting structure is the perfect sphere.

The subsymmetries described by the standard model are more complicated than simple circle symmetries, and the unifying GUT symmetry is more complicated than that of a sphere, but the basic

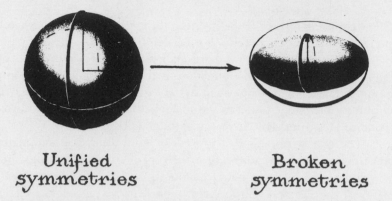

Unified symmetries

Broken symmetries

A perfect sphere has complete rotational symmetry and the radii of the circles can be thought to represent the strength of interactions, here both equal. If the symmetry is broken, the sphere squashes and its major and minor radii now might be thought to represent the different interaction strengths.

idea is similar. Viewed in this way, the strong, weak and electromagnetic interactions owe their parentage to a master GUT symmetry which is spontaneously broken. There are many ways that the symmetries of the standard model can be embedded into a larger single GUT symmetry, and guessing which way, and which larger symmetry, if any, is chosen by nature is the game played by modern model builders.

At first there seems to be a puzzle in unifying the strong colored gluons (which mediate the interactions among quarks) with the electro-weak gluons (which mediate interactions among both quarks and leptons). By unifying all these gluons under the aegis of one grand symmetry it seems we are also required to unify the quarks and leptons as well. If the GUT idea is to work, then all the quarks and leptons must be viewed as components of grand "lepto-quark" fields as well. From the standpoint of the unbroken single grand symmetry, quarks and leptons would be indistinguishable; they would transform into each other.

The physicists Jogesh C. Pati of the University of Maryland and Abdus Salam were the first to speculate about such a unification of the quarks and leptons. They suggested in 1973 that the leptons be viewed as a "fourth color" to be added to the three quark colors. When the four-color symmetry among the quarks and leptons then spontaneously broke down to the three-color exact symmetry of the strong quark-binding interaction, the quarks and leptons could be distinguished.

Their model had the further remarkable consequence that the proton could decay into lighter particles. The reason for this new feature of proton decay could be directly traced to the fact that in their models, quarks and leptons are unified as the components of a single lepto-quark field; therefore quarks, like the quarks inside a proton, can convert into leptons. Another way of stating this feature is that the baryonic-charge-conservation law that required proton stability in the standard model was violated in these new models. This was the first indication that most GUTs required proton decay to occur.

Unfortunately, Pati and Salam did not give a definite prediction for the proton decay rate (it was an arbitrary parameter of their model), nor did they provide a true unification of the strong and electro-weak interactions in the sense that all three interactions were manifestations of a *single* spontaneously broken gauge-field symmetry. But just one year later, in 1974, Howard Georgi and Sheldon Glashow at Harvard University devised the first model that implemented true unification—the "minimal SU(5) model"—which for its economy became the exemplar of all future GUTs. It is, as Georgi remarked, "very pretty."

The Georgi-Glashow model was the simplest of all GUTs. The 8 colored gluons and the 4 electro-weak gluons were incorporated as 12 components of a single 24-component Yang-Mills field; hence grand unification was achieved. This single grand symmetry was then spontaneously broken by Higgs symmetry-breaking fields. The result was that 8 of the original 24 gluons could be identified with the colored gluons, and 4 with the electro-weak W^+, W^-, Z and γ gluons. Twelve new gluons, called X gluons, became supermassive, so massive that no accelerator could possibly create them. Yet the new X gluons, through their interaction with the ordinary quarks and leptons, could convert quarks into leptons and vice versa. The supermassive X gluons therefore destabilized the proton. Once again grand unification implied proton decay.

Georgi and Glashow pointed out a further important implication of their GUT. The standard model, which does not unify the three interactions, has nineteen arbitrary parameters. One of these parameters, the "weak-interaction angle θ_w," was no longer arbitrary in the SU(5) model but was calculated to be equal to 37.7 degrees, a consequence of the exact unifying grand symmetry. This prediction was a step toward realizing Einstein's dream that "there are no *arbitrary* constants."

Unfortunately, their prediction disagreed with the value that the experimentalists had measured for this angle. Many people thought that SU(5) model, although it beautifully realized the GUT idea, was not a correct description of nature. But the idea of grand unification was becoming firmly entrenched in the minds of theorists, and they were hard at work exploring GUTs.

THE DESERT

Soon Howard Georgi, Helen Quinn and Steven Weinberg deepened the idea of GUTs by examining what they implied about the microscopic world at extremely short distances. GUTs were "renormalizable" field theories, which meant that the theorists could mathematically calculate the properties of quantum particles at very high energies or correspondingly very short distances. When this was done, a remarkable new view of the microcosm opened to the theorist's vision.

According to the renormalization idea, the physical properties of quantum particles, like their masses or interaction coupling strength, can change depending on the distance scale at which they are measured or observed. (Recall the analogy from a previous chapter to the fractal behavior of the length of a coastline.) What the theorists found is that if nature is indeed described by a GUT, then quantum interactions, in spite of their different strengths at long distances, would, when viewed at very short distance scales, reveal the exact underlying GUT symmetry. This meant that at ultrahigh energies, corresponding to those short distances, the strong, weak and electromagnetic interactions merged into a single unified interaction.

For example, suitably defined coupling constants for strong, weak and electromagnetic gluon interactions, denoted g_S, g_W and g_E respectively, although unequal for low-energy interactions (manifesting the broken symmetry), all became equal to each other at ultrahigh energy. This important point—that the underlying symmetries of a spontaneously broken field theory would be restored at very short distances—was originally emphasized by Kenneth Wilson through his novel work on the renormalization theory. His ideas now came to play an important role in the physical interpretation of GUTs.

These renormalization and distance-scale ideas were applied directly to the SU(5) model of Georgi and Glashow with remarkable consequences. We recall that this model unfortunately predicted the weak interaction angle θ_w incorrectly. But in the light of these new ideas, it became clear that this numerical prediction, based as it was on exact symmetry, applied only to the ultrashort distance scales at which the GUT symmetry became exact, not to the much longer distance scales observed in the laboratory and for which the angle was measured. So the problem now was to calculate the angle θ_w at the distance scales observed in the laboratory.

In the SU(5) model the distance scale for which the three interactions became equal—the so-called "GUT scale"—could be estimated through renormalization theory and was found to be 10^{-29} centimeters. This is an incredibly small distance if we realize that a proton is about 10^{-14} centimeters across and that it is already very small (though it is observable in modern accelerators). By using the renormalization theory, theorists could extrapolate the value of the weak interaction angle θ_w predicted at the GUT scale of 10^{-29} centimeters up to the distance scales of 10^{-14} centimeters observed in the laboratory. They calculated the value of the angle at these laboratory distance scales to be 27.2 degrees. Although an improvement, this value still did not quite agree with the experimental value. But as more experiments were done, the experimental value of the angle changed until it was 27.7 degrees—within measurement error, the value obtained by the theorists. The stock of the SU(5) model went up, and the theorists were convinced that they were on to something. The model became intensely investigated mathematically.

A curious picture of the microcosmic quantum world of small distances emerged from these investigations. The usual standard model tells us what happens down to a distance scale of 10^{-16} centimeters—the "electro-weak scale"—the size appropriate to the highest-mass particles which played a role in that model, the W and Z weak gluons. When the standard model is subsumed into the SU(5) GUT, the distinct interactions of the standard model are unified. Using the SU(5) GUT as a guide, theorists find that as the distance is further decreased from the electro-weak scale of 10^{-16} centimeters to the GUT scale of 10^{-29} centimeters, corresponding to the size of the X gluons, nothing much happens. No new particles appear, and the coupling strengths of the three interactions begin to smoothly approach equal values as the GUT symmetry begins to take hold.

Sheldon Glashow has appropriately dubbed this vast microscopic distance region "the desert" because it is empty of any new physics. As the GUT scale of 10^{-29} centimeters is approached, the X gluons merge with the other gluons of the standard model; all the gluon interactions are perfectly symmetrical and unified. Finally, at the incredible distance scale of 10^{-33} centimeters, the effects of quantum gravity (which have not been explicitly included in the GUTs) must become important. This is called the "Planck scale," after Max Planck, who first noticed that this was an important distance associated with gravity effects. Because the Planck scale is about one ten-thousandth the length of even the incredibly short GUT scale, most physicists feel comfortable ignoring the effects of quantum gravity in their speculations about GUTs.

According to these ideas, the microworld possesses a hierarchy of distance scales, milestones on the road to ever-shorter distances. The relatively low energies of today's particle accelerators enable physicists to explore down to the distance scales of the standard model, about 10^{-16} centimeters. At higher energies, physicists have calculated the other distance scales we have discussed, so that we have a hierarchy of distances:

$$
\begin{array}{ll}
\text{the electro-weak scale} & \simeq 10^{-16} \text{ centimeters} \\
\text{the GUT scale} & \simeq 10^{-29} \text{ centimeters} \\
\text{the Planck scale} & \simeq 10^{-33} \text{ centimeters}
\end{array}
$$

Between these microscopic distances nothing much happens—unlike the situation for much longer distances, for which there are complex particle interactions giving rise to hadrons, nuclei and atoms.

By taking ratios of the above three distance scales (so the fact that we chose centimeters to measure the distance drops out), we obtain the large pure numbers 10^{13} and 10^{17}. These numbers demand an explanation. What are they doing in nature? From the standpoint of GUTs or any other theory, these numbers, representing the hierarchy of distance scales in nature, are today without explanation. They are *"arbitrary* constants." Physicists want to explain these numbers and solve the "hierarchy problem." But so far, although there are some intriguing hints, a solution has eluded them. Even

GUTs have "*arbitrary* constants" and cannot be the final unified-field theories.

In spite of the fact that GUTs leave deep puzzles unsolved, they have gone a long way toward unifying the various quantum particles. For example, many people are disturbed by the large numbers of gluons, quarks and leptons. Part of the appeal of the GUT idea is that this proliferation of quantum particles is really superficial and that all the gluons as well as the quarks and leptons may be simply viewed as components of a few fundamental unifying fields. Under the GUT symmetry operation these field components transform into one another. The reason quantum particles appear to have different properties in nature is that the unifying symmetry is broken. The various gluons, quarks and leptons are analogous to the facets of a cut diamond, which appear differently according to the way the diamond is held but in fact are all manifestations of the same underlying object.

PROTON DECAY AND THE MATTER–ANTIMATTER ASYMMETRY OF THE UNIVERSE

Besides providing a definite conceptual picture of microscopic distances, the SU(5) model implied that the proton was unstable and had to decay. Theorists calculated the average lifetime for the proton in terms of other quantities that were already known and estimated that it would take the proton 10^{30} years to disintegrate into a positron and a neutral pion. This is 100 billion billion times the age of the universe—a very long time indeed. Nevertheless, this prediction provided a direct way to test the SU(5) model.

At the time the theorists made this prediction, experimentalists knew that the proton's lifetime exceeded 10^{29} years—the SU(5) model prediction of 10^{30} years was consistent with existing observations. The experimentalists were thus challenged by the theorists and the new GUTs to improve their observations on possible proton decay and test the prediction.

Searching for decaying protons is not a high-energy physics experiment to be done in one of the giant accelerators. Instead, what is required is a careful observation of a very large volume of matter—the larger the better—to see if any single proton in that matter disintegrates. The search for proton decay exemplifies a different kind of experiment done to check the fundamental laws of physics, one that tends away from high-energy experiments toward very-low-energy but large-volume experiments. Even while experimental physicists were outfitting sensitive detection systems in large volumes of matter to look for possible proton decays, the theorists were speculating on the significance of proton decay for the largest volume in existence— the entire universe. Here they discovered something quite remarkable.

Recall that in my description of the big bang, one of the inputs to our supercomputer was the specific entropy—the ratio of the number of photons to the number of baryons (protons and neutrons)—a number about 400 million to 1. The tiny number of baryons corresponds to all the visible matter in the universe today. The puzzle posed by this small number of baryons is Why isn't it zero; why did such a small but finite number of protons and neutrons survive the big bang? This puzzle is compounded by the fact that the law of baryon-number conservation means that this net baryon number has to have been present from the very beginning of the universe. That seems a rather arbitrary initial condition.

A far more attractive initial condition is perfect symmetry between baryons and antibaryons, effectively a zero net baryon number at the instant of creation. But if baryonic number is conserved, then the total number of baryons would equal the number of antibaryons and we would live in a world of matter–antimatter symmetry. But the observed universe does not have such a matter–antimatter symmetry —it is made mostly of matter—and hence beginning with such a symmetry is a problem.

GUTs solve this problem. In most GUTs baryon number is not conserved and there is the exciting possibility that the universe could have begun in a matter–antimatter symmetric state and then gone on to create its own matter–antimatter asymmetry. The possibility of proton decay in GUTs—the prime example of a baryon-number-violating process—also implies that protons can be created, and provides the clue to answering the question of why the visible universe exists.

In 1968, even before GUTs were invented, Andrei Sakharov, a Soviet scientist, realized that if baryon number is not conserved, it would explain in part how the matter–antimatter asymmetry we see today could have arisen from a state of perfect symmetry. Sakharov also realized that baryonic-number nonconservation, while a necessary condition for the creation of matter, was not sufficient. Other conditions had to be met.

The first of these conditions is that the universe has to make more matter than antimatter. In order for it to do this, matter and antimatter, which are through-the-looking-glass versions of each other, have to be distinguished by some interaction that tells us on which side of the looking glass the present universe is. Experimentalists have actually detected such interactions (they are called time reversal-violating interactions), so this condition *is* met.

Another condition is that the universe must, during a very early stage of its development when the baryon-number-violating processes are most effective, be in a state of nonequilibrium. This means that at some time in its early history the universe must undergo a "phase transition," a change of its basic state which happens so rapidly that the rate of collisions between the quantum particles in the primordial gas cannot keep up with it. If such a phase transition occurs, any matter–antimatter asymmetry that gets generated during the transition also gets to stay, because once the transition is over, the baryon-number-violating processes become less effective and baryon-number conservation is effectively restored. The "extra" protons and neutrons, generated out of nothing, are now locked into the universe.

In summary, a matter–antimatter asymmetry could be generated starting from a symmetric state provided that 1, baryon number was not conserved; 2, time-reversal-violating interactions exist; 3, the universe was once in a nonequilibrium state of extreme expansion. In the standard model neither condition 1 nor 3 is met, and hence the origin of visible matter remains a puzzle in this model. However, GUTs, which go beyond the standard model, can violate baryon-number conservation. This led to a revival of interest in explaining the observed matter–antimatter asymmetry. By 1978 many theoretical physicists, realizing that this old puzzle could now be explained in the context of GUTs, were hard at work calculating the asymmetry to see if it would agree with observation.

A few years ago a T-shirt with the slogan "COSMOLOGY TAKES GUTs" enjoyed modest popularity among cosmologists. The slogan meant that GUTs could solve the problem of the origin of the matter–antimatter asymmetry and thus explain the genesis of the visible universe. Yet another meaning of the slogan is that many

GUTs (the SU(5) model is an exception) imply that the neutrinos are not strictly massless and hence could make up the dark matter of the cosmos. GUTs might provide the answer to the origin not only of visible matter but of the invisible matter as well.

Using the SU(5) model, theoretical physicists estimated the number of nuclear particles (the specific entropy) that were made in the very early universe and got a number that was too small by factors of 10 to 100. However, this estimate depended on details of the early universe when it was only 10^{-35} second old and that were not well understood. So the fact that the estimate was off by such large factors was viewed not as a failure but rather as a sign that the new ideas were actually working. Subsequently, with the construction of other GUT models, the numbers came out right. Just as physicists in the late 1960s calculated the big-bang genesis of elements like helium, deuterium and lithium out of protons and neutrons, the physicists of the late 1970s were calculating the genesis of baryons like protons and neutrons out of quarks and leptons. More and more physicists were banking on the GUTs, especially the SU(5) model.

Meanwhile, experimentalists were improving the measured limit on the proton lifetime, which they knew exceeded 10^{29} years—a very long time. How can experimentalists expect to limit the lifetime of something which already exceeds the lifetime of the universe by 10 billion billion years?

Maurice Goldhaber, a physicist at the Brookhaven National Laboratory who has thought a lot about possible proton decay, once remarked that "we know in our bones" that the average lifetime of the proton exceeds 10^{16} years. The human body contains about 10^{28} protons; if the proton lifetime were smaller than 10^{16} years, this would correspond to 30,000 decays a second and your own body would be a radioactive health hazard. As even this rough estimate suggests, more stringent limits on the proton lifetime can be obtained if we use a volume with more protons than our bodies and a better detection system than our state of health.

Fortunately, there are many protons around. According to the quantum theory, protons, if they decay at all, must decay at random. This implies that if physicists observe a sufficiently large number of protons, the proton lifetime in years is given simply by the total number of protons under observation divided by the number of proton decays actually observed in one year. Hence the main limitation on measurement of the proton lifetime is the limit on the total number that can be carefully observed and the detection efficiency for actually spotting proton decay, should it occur.

In order to minimize background events from cosmic rays (which might be mistaken for proton decay), thereby enhancing detection efficiency, physicists have situated experiments to observe proton decay deep underground. Experiments are being conducted in the Soudan mine in Minnesota, the Mont Blanc tunnel between Italy and France, the Kolar gold field in southern India, the Caucasus range of the Soviet Union and the Silver King Mine of Utah. One such experiment, a collaboration between the University of California at Irvine, the University of Michigan and Brookhaven National Laboratory, utilizes 8,000 tons of water—a huge swimming pool—1,950 feet underground in the Morton salt mine east of Cleveland, Ohio. Throughout the volume of water are sensitive photocell detectors for the telltale products of a single proton decay (a neutral pion π° and a positron e^+) out of all those in the water. In late 1982 the experimentalists reported that no such events had been seen and that the proton lifetime into the $\pi^\circ + e^+$ decay mode must therefore exceed 10^{32} years. This implied that the simplest SU(5) GUT model, which estimated the lifetime to be about 10^{30-31} years, is wrong. This was a disappointment to many people, especially the theorists. But in spite of that disappointment the idea of GUTs is being pursued.

The theorists have lots of other GUTs that can accommodate the new experimental limits on the proton lifetime. Some of these models imply that the dominant products of proton decay are not π° + e^+ at all, but entirely different particles which are harder to see. But as physicists continue to run their experiments, the limits will improve and even some of these more elaborate models might get ruled out.

If proton decay is eventually seen, it will be enormously exciting. By studying the decay, physicists could obtain more detailed information which would be very useful in determining which, if any, GUT model describes nature. Strange as it seems, swimming pools of water deep in a mine may provide clues about the origin of the universe.

BEYOND SIMPLE GUTs

Many physicists are suspicious about the assumption that nature chooses a simple GUT. They are especially uneasy about a model that correctly describes nature at distance scales of 10^{-16} centimeters (the scale size of the W and Z weak gluons) and then extrapolates it to distances of 10^{-29} centimeters (the GUT scale). This extrapolation is 13 powers of ten—approximately the thickness of a finger compared with the distance between the earth and the sun. If the simple GUT idea is right, then nature has a microdesert covering an immense region. Many physicists would view that as an absence of imagination on the part of nature.

Scientists observed that from the macroscopic scale of superclusters of galaxies to the microscopic scale of the W and Z gluons, nature reveals new physical structures as the distance scale changes. In Philip Morrison's book *Powers of Ten*, we see illustrated the richness and variety of nature's productions every time the distance scale changes by a factor of ten. Is there some reason (other than lack of imagination or data) for this richness to suddenly cease as we probe beyond distances corresponding to the size of the W and Z gluons only to become interesting again at the GUT scale—13 powers of ten smaller? No one knows the answer to this question. The only way to find the answer is to continue doing accelerator experiments at still higher energy.

Physicists are planning to build several large accelerators. The Europeans at CERN near Geneva have begun the construction of LEP, a machine in which electrons and positrons are accelerated up to high energy in two counterrotating circular beams (the electrons moving in one direction, the positrons in another) which then collide. By studying the debris of the collision, physicists hope to find new particles like the Higgs particles or new heavy quarks and leptons.

Physicists in the United States are also planning a supercollider accelerator to dwarf all others. It is nicknamed the "desertron"—not only because it can explore the GUT desert but also because the machine is so large it must be built in the American desert.

A few years ago during a visit to the Los Alamos Scientific Laboratory in New Mexico, I stopped in to see a colleague who was searching out possible locations for this monstrous machine. On his office wall were maps of the American desert spotted with hand-drawn circles (the shape of the machine) indicating possible sites. The

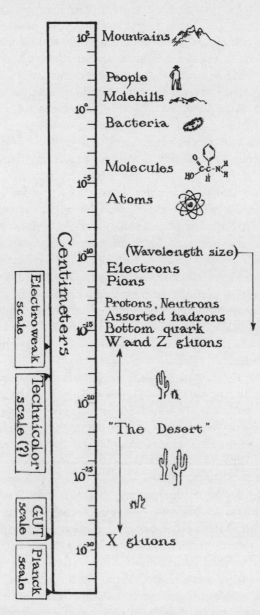

The distance scales of the universe in powers of 10, from the Planck scale to the size of the universe. There may be a "desert" in the microworld. One of the main insights of modern physics is how the microworld influences the macroworld.

accelerator might be as large as 30 miles in diameter, so a rather large flat region of desert had to be found to minimize excavation costs. After physicists complete their deliberations on what design best suits the purpose of exploring the unknown realms of the quantum particles, they will make a proposal to the government to build it. The energies to be reached in such a machine will be comparable to those in the gas of quantum particles when the universe was fractions of a nanosecond old.

Even while such accelerators are being designed, theoretical physicists, unhampered by such down-to-earth problems, are busy constructing models of what might be discovered at the desertron collider—an imaginative enterprise known as "populating the desert." Some GUTs like the SU(5) model imply that only a vast microworld of desert exists beyond distances of 10^{16} centimeters and predict that nothing new will be found by the supercollider. But a feature of all such simple GUTs as the SU(5) model is that they unify only *three* known forces of nature—the strong, weak and electromagnetic forces. Perhaps completely new forces exist which are distinct from these three and gravity, but are manifested only at very short distances. Such forces will not be revealed until accelerators with the energy capable of probing them are built—energies that are available at the supercollider.

One such new force, which so far exists only in the imagination, was independently conjectured by Steven Weinberg of the University of Texas and Leonard Susskind of Stanford University and dubbed the "technicolor" or "hypercolor" force. These theorists guess that it might show up at distance scales of about 10^{-17} or 10^{-18} centimeters. The technicolor force imitates the usual color force that binds quarks together. But it is mediated by a new set of "technicolor gluons" that interact with a new set of quarks, the "techniquarks," which bind together to form "technihadrons"—similar to ordinary hadrons like the protons, neutrons and pions except that they are very much more massive. If these ideas about technicolor are right, then new technihadrons will be created in the mighty new accelerators in just the way ordinary hadrons were created in the old accelerators.

Physicists did not introduce the technicolor force just to imagine hypothetical new particles that could populate the desert. They were also hoping to deepen their understanding of unexplained parameters in the standard model such as the magnitude of the ordinary quark masses. Perhaps the new technicolor forces could explain the observed quark masses. Unfortunately, the idea of technicolor forces has not been very successful in elucidating such unsolved problems, and it runs into further problems if one tries to combine the usual weak interactions with the technicolor forces. In spite of these difficulties, technicolor is a wild idea that might turn out to be important when the new machines turn on. No one can rule out the existence of new, very-short-range forces, and the quantum particles associated with those forces could make the desert bloom.

GUTs, and the technicolor forces that can be added to them, represent but one (albeit the major) attempt to unify the forces of nature. Another suggestion to unify the various quantum particles is that they are not elementary but composite entities. In the past, whenever physicists found what they took to be an "elementary" object like the atom or the proton, they subsequently discovered that the object was in fact composed of yet smaller objects. Why aren't the quarks, leptons and gluons subject to further decomposition?

Maybe they are composite. Yet physicists who have explored this suggestion have not yet found a way to apply it. In the past, the assumption that a particle was composite (in the absence of any direct experimental evidence for compositeness) usually explained some otherwise puzzling property of that particle. For example, the assumption that atoms were made of electrons and nuclei helped explain the spectrum of light they emitted. Likewise, the assumption that hadrons were composed of quarks implied correct relations among the observed hadron masses.

Pati and Salam in their work on unification explored the idea of "preons"—smaller objects out of which the quarks and leptons might be built. The Israeli physicist Haim Harrari developed a "rishon" model of the quarks and leptons in which all the observed quarks and leptons may be viewed as built out of two rishons. But in spite of these and many other efforts, the idea that quarks and leptons are composites has not been helpful in explaining masses or other properties of quarks and leptons. That is discouraging. Perhaps it is because physicists have not yet found the right composite model or are not applying the ideas of compositeness properly. The notion of composite quarks and leptons is yet another wild idea to be put on the shelf until someone figures out how to make it work.

Even though there is not one piece of experimental data that directly supports them, GUTs are at the forefront of theoretical physics research today. Physicists are encouraged by the fact that GUTs enable them to unify three forces of nature, at least in theory. Two decades ago such a unification would have been unthinkable. Today the unification of all the forces of nature is on the mainstream agenda of modern physics.

Finding evidence of proton decay will be an important confirmation of GUTs. But even if experimentalists do not see decaying protons, this only means that the proton lifetime may be longer than they can measure. The absence of proton decay would then provide another datum constraining the theorists' models, not the end of the GUT idea.

Many theorists today are working to further generalize GUTs to include a new kind of symmetry—supersymmetry—which involves transforming half-integer-spin fields into integer-spin fields and vice versa. These models go by the name of "SUSY GUTs"—SUper-SYmmetric Grand Unified Theories. The proving ground for these SUSY GUTs is the very early universe—the dynamics of the inflationary universe, a postulated stage of the universe prior to the standard big bang. Other, even grander theories than GUTs are the supergravity theories, extensions of Einstein's general theory of relativity that bring in the fourth force, gravity, and are the topics of a subsequent chapter.

The idea of GUTs grew out of the success of the gauge-field-theory revolution and the subsequent standard model of quarks, leptons and gluons. But GUTs were not the only product of this scientific revolution. Physicists mathematically investigating gauge-field theories discovered a whole new class of objects that might inhabit the quantum microworld. The most exciting of these is the magnetic monopole, a particle possessing a single unit of magnetic charge and which is unlike anything ever seen before. Magnetic monopoles are predicted by many GUTs, and if these theories are right, then we may discover monopoles someday. The intriguing story of magnetic monopoles follows.

2

Magnetic Monopoles

From the theoretical point of view one would think that monopoles should exist, because of the prettiness of the mathematics. Many attempts to find them have been made, but all have been unsuccessful. One should conclude that pretty mathematics by itself is not an adequate reason for nature to have made use of a theory. We still have much to learn in seeking for the basic principles of nature.

—P. A. M. Dirac, 1981

When Albert Einstein was four his father gave him a magnetic compass—a small bar magnet free to pivot—which behaved in a wonderful way that "did not at all fit into the nature of events which could find a place in the unconscious world of concepts." Einstein's compass was responding to the earth's invisible yet detectable magnetic field. This experience stimulated his interest in physics, and later Einstein went on to make major contributions elucidating the nature of magnetism and electricity. Yet he, like other physicists, was puzzled by an unusual asymmetry between magnetism and electricity—there are no magnetic charges comparable to electric charges. Our world is filled with electrically charged particles like electrons or protons, but no one has ever detected an isolated magnetic charge. The hypothetical object that would possess it is called a magnetic monopole.

To get an idea of what a magnetic monopole would be like if one existed, imagine a bar magnet like a compass needle, with a north and a south pole at its ends. The magnetic field of a bar magnet can be visualized as lines of force exiting the north pole and reentering the magnet at the south pole and then being channeled through the bar back to the north pole. If you sprinkle iron filings in the vicinity of the magnet, you can actually see those lines of force in nature. Such a field configuration is an example of a "dipole field"—it has two poles, the north and the south pole, and its field lines never end: they loop around endlessly. If the bar magnet is cut in half, the result is not separate north and south poles but two bar magnets. To find an isolated north or south pole—an object with magnetic-field lines only exiting or only entering—would be to discover a magnetic monopole. For reasons not clear, nature either did not create magnetic monopoles or created very few of them.

By contrast, electric monopoles—particles that carry electric charge—are abundant. Every speck of matter contains incredible numbers of electrons and protons, true electric monopoles. One may visualize the electric-field lines of force emerging from or converging on an electrically charged particle and beginning or ending there. Furthermore, experience has confirmed the law of electric-charge conservation: the total electric monopole charge of a closed system can be neither created nor destroyed. But nothing similar to electric monopoles exists in the world of magnetism, in spite of the fact that a magnetic monopole is easily imagined.

James Clerk Maxwell, the Scottish physicist who mathematically unified the electric and magnetic fields in 1864, included in his fundamental electromagnetic equations the existence of electric charges but did not include possible magnetic charges. It would have been easy for him to do so; aesthetically, such an inclusion would have made his equations beautifully symmetrical with respect to electricity and magnetism. Yet like other physicists of his time, he saw no evidence for magnetic charges in nature and by fiat banished them from his equations. The natural asymmetry of electricity and magnetism has struck physicists as peculiar ever since.

Physicists continued to deepen their understanding of the electromagnetic Maxwellian field. They knew that Maxwell's equations could be simplified if the electric and magnetic fields were mathematically derived from a yet more fundamental field—a gauge field. The electromagnetic gauge field is the first and simplest example of the general gauge-field concept discovered by Yang and Mills much later. Interestingly, by requiring that Maxwell's equations be expressed in terms of the simple gauge field, physicists saw that the absence of magnetic charge was now mathematically accounted for. Conversely, they could show that the absence of magnetic charge mathematically implied the existence of a gauge field. The gauge field thus introduced an asymmetry between the electric and magnetic fields.

But the introduction of the gauge field as the underlying structure of electromagnetism seemed at the time like a mathematical novelty —a conceptual trick and not real physics. One got out of the gauge-field idea—no magnetic charge—precisely what was put in—no magnetic charge. Then, in the 1920s, the mathematician Hermann Weyl showed that incorporation of electric and magnetic fields into the new quantum theory actually required a gauge-field description. That is when physicists began to see that the electromagnetic gauge field was physically important as well as mathematically interesting. Quantum mechanics seemed "made" for gauge fields, and remarkably, gauge fields implied the absence of magnetic monopoles. This theoretical outlook accorded so completely with experience that the electromagnetic-gauge-field concept took firm hold. Then came Paul Dirac.

In 1931, Dirac set out to examine the physical implications of the "pretty mathematics" of the electromagnetic gauge field in the quantum theory. He remarked that "When I did this work I was hoping to find some explanation of the fine structure constant [the constant related to the fundamental unit of electric charge]. But this failed. The mathematics led inexorably to the monopole." Contrary to the prevailing theoretical outlook, Dirac found that the existence of an electromagnetic gauge field and quantum theory together implied that magnetic monopoles could in fact exist—*provided* that the fundamental unit of magnetic charge had a specific value. The value of the magnetic charge Dirac found was so large that if such magnetic monopoles exist at all in nature they would be easily detectable through the effects of their large magnetic fields.

One way to visualize Dirac's result is to imagine a thin mile-long bar magnet with a magnetic field emerging from each end. Here, the magnetic field resembles that of a magnetic monopole because the magnet is so thin and the ends are so far away. But it is not a true monopole because the magnetic-field lines do not actually end at the tip of the thin magnet; they are channeled through the magnet and emerge at the other end.

Next imagine that one end of this thin magnet extends into infinity and that the thickness of the bar magnet is mathematically reduced to zero. The magnet now resembles a mathematical line or string with a radial magnetic field emerging from its tip—a true point magnetic monopole. But what about that infinitely thin string (called a Dirac string) which channels the magnetic-field flux out to infinity? Remarkably, Dirac proved that provided the magnetic charge of the monopole, with a value g, satisfied the equation

$$ge = n/2 \qquad n = 0, \pm 1, \pm 2, \ldots$$

where e is the fundamental unit of electric charge (an experimentally known quantity), the presence of such a string could never be physically detected. According to Dirac, the string then becomes simply a mathematical descriptive artifact without physical reality, just as coordinate lines on maps are mathematical artifacts of our description of the surface of the earth and have no physical significance. Mathematically, the Dirac string with a magnetic monopole at its tip was a line in space along which the electromagnetic gauge field was not defined. But, amazingly, this lack of definition had no measurable consequence, provided that the magnetic-monopole charge satisfied Dirac's condition. A further consequence of Dirac's monopole was that magnetic charge, like electric charge, would be rigorously conserved.

A colleague of mine once heard Dirac lecture on the magnetic monopole. To illustrate the mathematical string, Dirac produced a real string and, holding one end fixed—representing the location of the magnetic monopole—he moved the rest of the string about, asserting that its orientation had no physical consequences. After the lecture my colleague privately asked Dirac if he kept the string for the express purpose of this lecture demonstration. Dirac enigmatically responded to the effect "No, I had that string in my pocket long before I began thinking about magnetic monopoles."

After Dirac's important work, theoretical physicists accepted the possible existence of magnetic monopoles, reasoning that if no law of physics forbids their existence then perhaps they exist. Experimental physicists were encouraged to search for them.

Maybe magnetic monopoles are embedded in ordinary matter and could be extracted with strong magnetic fields. Other physicists tried producing them at particle accelerators, but none were found. The search for monopoles extended to cosmic rays—showers of quantum particles and atomic nuclei raining down upon the earth. On one exciting occasion, a track on a photographic emulsion exposed in the upper atmosphere by scientists at the University of California at Berkeley looked as if it might be that of a magnetic monopole. But most people are now convinced that that particular track was made by a heavy nucleus plowing through the emulsion, a track that could easily be confused with one expected for a monopole.

An effective modern technique for searching for magnetic monopoles consists of using a superconducting ring—a closed electrical conductor with no resistance to electric current. An electric current set up in such a ring persists forever. If such a superconducting ring has exactly zero electric current and if a magnetic monopole were to fly through the ring, then an electric current would start to circulate and this current would be easily detected. Furthermore, the strength of that induced current is precisely related to the strength of the magnetic-monopole charge.

Such a superconducting ring was set up by Blas Cabrera at Stanford University. On February 14, 1982, at nearly 2 P.M., the current in the ring jumped from zero to precisely the value expected if a monopole had gone through the ring. Physicists again became excited.

The event was never repeated. Perhaps the current *was* caused by a monopole and monopoles may be so rare that this was a lucky catch. However, physics is based on reproducible results. Until the experiment can be repeated, most physicists will think of the event as a fluke. The experimental search for magnetic monopoles continues—"but all have been unsuccessful."

After the gauge-field revolution of the early 1970s, physicists created a whole new understanding of magnetic monopoles which went well beyond Dirac's earlier work. The innovators of the new theory were Gerard 't Hooft, a young Dutch physicist who had already made major contributions to the theory of gauge fields, and A. M. Polyakov, a brilliant young Soviet physicist. In 1974 they mathematically demonstrated that some of the new Yang-Mills gauge-field theories, if their gauge-field symmetry was spontaneously broken—as was often the case—produced mathematical solutions for the field configurations that corresponded to magnetic monopoles. Furthermore, physicists could calculate the detailed properties of these mono-poles—their mass, spin and so forth. This was unlike the Dirac monopole, for which such properties were left unspecified.

Prior to these discoveries, many theoretical physicists thought that the solutions to field theories only described the particles correspond-ing to the fields in the equations. Yet here was an entirely new class of unanticipated solutions to field theories identified as "topological solitons" that corresponded to magnetic monopoles. In order to de-scribe these surprising new riches of gauge-field theory, I must say a few words about the mathematical concept of "topology" as well as what is meant by a "soliton."

Topology, a highly developed branch of pure mathematics, deals with the unchanging properties of mathematical objects—such as geometrical figures—that do not depend on how the object is contin-uously transformed. Imagine the surface of a perfect sphere and imagine deforming this surface into an ellipsoid or a long cigar or any other shape. Next imagine the surface of a doughnut and arbi-trarily deform that surface in your mind. No matter how you deform the doughnut, there is no way that you can continuously transform it into a sphere without tearing up its surface so that two nearby points on the surface become discontinuously separated. You just cannot get rid of the hole in the doughnut. This simple example illustrates the fact that the surfaces of a sphere and of a doughnut are topologically distinct—there is no way to continuously transform one shape into the other.

As a second example of two topologically distinct configurations, imagine a vector field—lots of little arrows—in a two-dimensional plane. Each point in the plane is associated with a vector—a magni-tude and a direction—and the set of all these vectors is called a vector field. Consider the two different vector-field configurations shown in the illustration with the vector fields represented by little arrows, each arrow of the same length and pinned to a point. Can these two field configurations be continuously transformed one into the other by rotation of each arrow about its point? If the transformation is to be continuous (analogous to not tearing the surface of the sphere or doughnut in the previous example), then this implies that the amount we rotate each arrow in the plane can differ only infinitesimally from the amount we rotate infinitesimally nearby arrows. If we try align-ing the arrows in the B configuration so that it resembles the A

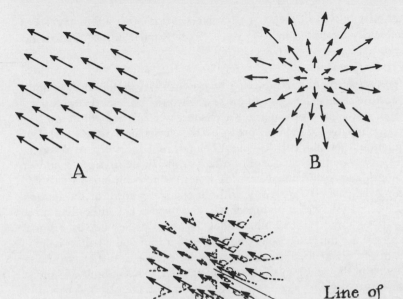

A

B

Line of
discontinuity

The "vacuum" configuration, A, and the "hedgehog" configuration, B, of field arrows. These configurations are topologically distinct. If one tries to convert configuration B into configuration A by rotating the field arrows, one finds that this cannot be done with only small differences in the rotation of neighboring field arrows; a line always separates a large difference of rotation of neighboring arrows. In the case of the monopole, such a line can be identified with a Dirac string.

configuration with all the arrows pointing in the same direction, we find that try as we will, there is always a line of discontinuity in the plane (an example is shown). On one side of the line the arrows are rotated clockwise and on the other side, counterclockwise. Since such a line represents a discontinuous jump in the sense of rotation of adjacent vectors, we conclude that the A- and B-field configurations cannot be continuously transformed one into the other; the field configurations are topologically distinct like the sphere and the doughnut. These examples of topologically distinct configurations suffice for our topology lesson. Next we ask: what is a soliton?

The first example of a soliton is a solitary water wave—a lump of

water that moves of its own peculiar accord. Such a solitary wave
was vividly described in 1844 by J. Scott-Russell in his "Report on
Waves":

I was observing the motion of a boat which was rapidly drawn along
a narrow channel by a pair of horses, when the boat suddenly stopped
—not so the mass of water in the channel which it had put in motion;
it accumulated round the prow of the vessel in a state of violent
agitation, then suddenly leaving it behind, rolled forward with a great
velocity, assuming the form of a large solitary elevation, a rounded,
smooth and well-defined heap of water, which continued its course
along the channel apparently without change of form or diminution
of speed. I followed it on horseback, and overtook it still rolling on at
a rate of some eight or nine miles an hour, preserving its original
figure some thirty feet long and a foot to a foot and a half in height.
Its height gradually diminished, and after a chase of one or two miles
I lost it in the windings of the channel. Such, in the month of August
1834, was my first chance interview with that singular and beautiful
phenomenon.

In 1895, two Dutch mathematicians, Diederick Johannes Korteweg
and his student Gustav de Vries, developed a mathematical equation
intended to explain the water wave chased by Scott-Russell. This
work initiated the mathematical study of solitons, as the solutions to
their equation were called, and many other examples were extensively explored by mathematicians in subsequent decades. Although
physicists knew of this mathematical work and its applications to
water waves, or waves in electrical plasmas of charged particles, they
did not suspect that soliton solutions were lurking in the equations of
some of their favorite Yang-Mills field theories. But there they were.

These new solutions to the equations of field theory are best described as lumps of bound field energy the way the first soliton was
a lump of water. These localized lumps look just like other quantum
particles, but unlike other quantum particles—the quarks, leptons
and gluons—they do not correspond to quanta of a fundamental
field. Instead they are built up out of a curiously twisted configuration of the fundamental fields. These twisted lumps of field energy,
the solitons, are the magnetic monopoles discovered by 't Hooft and
Polyakov.

If a soliton is just a lump of field energy, why doesn't it fall apart
and dissipate into its constituent fields? What holds the 't Hooft-

Polyakov soliton together? To answer this question we invoke topology. Some of the fields in the soliton are just like the vector field I described in the two-dimensional plane except that they are in real three-dimensional space directed radially outward from a point. Polyakov called such a field configuration a "hedgehog" because the vectors all point outward like the needles on the back of a hedgehog in its defensive posture. The other field configuration in which all the vectors are aligned in the same direction corresponds to the "vacuum"—it represents the absence of fields. The "hedgehog" and the "vacuum" are topologically distinct configurations, and to convert one into the other necessitates that we "tear" the fields apart. This requires infinite energy, and since infinite energy is not available, a single "hedgehog" soliton cannot dissipate into the "vacuum." We learn that it is the topological features of the soliton's field configuration which guarantee its stability. It is a lump that lives forever.

The only way to destroy or create such a topological soliton is to bring it together with a topological antisoliton. In the antisoliton the hedgehog vectors point radially inward; if one lays such a field configuration upon the outward-pointing soliton, the two configurations just cancel and can be continuously transformed into the vacuum. Such topological solitons can be thought of as possessing a "topological charge" which is absolutely conserved—a charge of plus one unit for the soliton and minus one for the antisoliton. The conservation of this topological charge guarantees the stability of the soliton.

The solutions to the gauge-field theories discovered by 't Hooft and Polyakov had all these marvelous topological properties. Amazingly, all that remained of the fields far away from the lump was a radially directed magnetic field. The lump, a topological soliton, was a magnetic monopole! The topological charge of the soliton was precisely the magnetic charge of the monopole. With the work of 't Hooft and Polyakov, the monopole, relegated to the status of a curiosity for decades, returned to physics with a vengeance (as we will shortly see). Ever since this theoretical work on the monopole, other topological solitons called instantons, vortices and kinks came to play an increasingly important role in the understanding of field theory.

The 't Hooft–Polyakov magnetic monopole is related to the earlier Dirac monopole. We saw that attempting to align all the hedgehog vectors in the same direction by a continuous transformation was impossible. There was a line ending at the central point along which the transformation was not continuous. This singular line is precisely the Dirac string which was required in Dirac's treatment of the magnetic monopole. It pops up as soon as we try to transform a hedgehog into a vacuum configuration. Because it takes infinite energy to get rid of the Dirac string, we again conclude that the monopole must be absolutely stable. You just cannot twist a doughnut into a sphere.

Some of the new gauge-field theories explored by the theorists possessed these new topological solitons corresponding to magnetic monopoles and others did not. By application of the topology lessons of pure mathematicians, it was easy to determine which theories had the monopole solutions and which did not. The standard model, which most physicists accepted as an adequate description of nature for distances larger than 10^{-16} centimeters—the scale size of the W and Z gluons—did not have monopole solutions. But the moment physicists went beyond the standard model by attempting to build models that unify the strong color force with the electro-weak force as in GUTs, they found that the topology of field configurations allowed for lots of monopoles. These monopoles, because they occur in GUTs, are called "GUT monopoles."

It is easy to estimate the mass of a GUT monopole by examining the properties of the topological-soliton solutions corresponding to the monopole. The mass is about one hundred times the GUT mass scale (10^{15} times the proton mass)—about a microgram. This is an immense mass, far larger than the mass of even the smaller macroscopic objects like bacteria. Such GUT monopoles, if they exist, would be wonderful, exotic objects never before seen.

Because of their immense mass, it takes immense energy to produce GUT monopoles. No existing or planned particle accelerator could possibly do it; even the energy release in a supernova is insufficient to create such magnetic monopoles. But there was one time when they could have been produced because the energy was available: the beginning of the big bang. Theorists, beginning with Yakob B. Zel'dovich and M. Yu. Khlopov in the Soviet Union and John Preskill at Harvard University, calculated the number of GUT monopoles produced in the big bang and found the number to be immense. According to these calculations, most of the mass of the universe today should be in the form of magnetic monopoles—an absurd result. Here in their favorite GUTs theorists found an absurd result.

This, then, was the vengeance of the magnetic monopole. Monopoles, once considered mere curiosities, were now forced on physicists thinking about the early universe. According to their calculations, there should be many of them—far too many. Where are they?

Terrestrial searches came up with nothing. But why restrict the search for magnetic monopoles to finding them on earth? If magnetic monopoles exist, they should be flying around in interstellar and intergalactic space along with the other debris left over from the big bang. Cosmic-ray physicists have looked for monopoles in the cascades of particles raining down on the upper atmosphere each day and found none. But if physicists look beyond the cosmic-ray fluxes of particles falling on the earth deep into the galaxy, they obtain a more severe restriction on the number of magnetic monopoles in the universe.

Our galaxy has a magnetic field that winds through the spiral arms, and the intensity of this immense field has been measured by its influence on particles and light. Astronomers assure us that this field has existed for a very long time—certainly 100 million years if not for the entire lifetime of the galaxy. The existence of the galactic magnetic field severely constrains the existence of magnetic monopoles—for if monopoles existed in any great numbers, they would long ago have eaten up the galactic field and there would be no such field today.

Magnetic monopoles are accelerated by a magnetic field just as electrically charged particles are accelerated by electric fields. As magnetic monopoles are accelerated, they gain energy which they have acquired from the magnetic field. As a consequence of its energy loss, the magnetic field reduces its intensity and soon it is gone. We see that the presence of magnetic monopoles in any great numbers in outer space is inimical to the long-term existence of the galactic magnetic field. Hence, the fact that such a field exists today imposes a severe bound on the density of monopoles—a limit known as the"Parker bound" after the physicist who first established it. If the Parker bound is valid, then one can show that any terrestrially based experiments like those done with superconducting rings will never "catch" magnetic monopoles—they are just too rare. The Parker bound, which uses the observed strength of the galactic magnetic field, is yet another example of how astronomical observations constrain the wild models of the theoretical physicists.

GUT magnetic monopoles are extended objects—they can be visualized as tiny little balls of field energy. Imagine approaching such a monopole. From far away all one sees is its magnetic field, but as one approaches its "surface" one begins to detect particle–antiparticle pairs being created and annihilated by the intense magnetic field. Once inside the monopole one encounters every kind of interacting quanta—the leptons, quarks, and colored gluons. But as one penetrates further a curious thing happens: broken gauge symmetries become restored. For example, penetrating to a distance of 10^{-16} centimeters from the center of the monopole—the distance scale of the electro-weak unification—one finds that the electro-weak symmetry is restored. At the very core of the monopole, which begins at a radius of 10^{-29} centimeters from the center—the GUT scale—all symmetry is completely restored. Here in the tiny core of the monopole we find a state of matter with complete symmetry resembling the state of the very early universe. The amazing GUT monopoles carry inside them the entire thermal history of the universe. They are

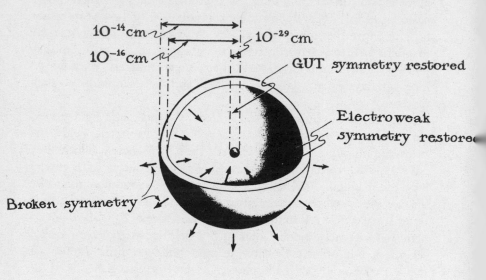

A GUT monopole: X-section

A GUT monopole reveals the entire thermal history of the universe. As one proceeds to the center of the monopole, broken symmetries are restored. Outside the monopole is the world of broken symmetry. Just inside, the electro-weak symmetry is restored, and at the very core, about 10^{-29} centimeters across, the full GUT symmetry is restored.

amazing onions, with each layer corresponding to an era of the early universe. This history goes all the way back to the time they were created, the time in the early universe when the temperature was so high the GUT symmetry was exact.

The exact symmetry at the core of the GUT monopole has implications for proton decay. There is a definite but small probability that a proton colliding head on with a monopole will touch the core. If this happens, the symmetrical GUT interactions which violate proton-number conservation can cause the proton to quickly decay into lighter particles. The monopole is unscathed by the collision, since its magnetic charge is absolutely conserved. Monopoles thus catalyze proton decay—a process known after its discoverers as the Rubikov-Callen effect.

If a GUT monopole were to fly through the swimming pool of 8,000 tons of water outfitted to observe proton decays, it would leave a trail of decaying protons in its wake. Such a spectacular event would signal not only proton decay but the existence of monopoles.

If monopoles catalyze proton decay, then some physicists are con-

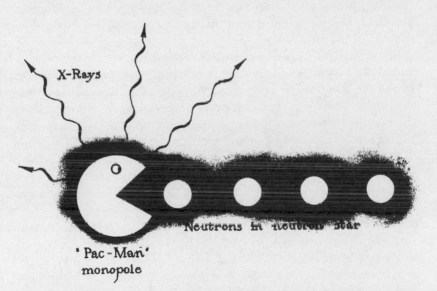

Pac-Man eats video dots the way a GUT monopole eats neutrons inside a neutron star. This process generates X rays. From the measured limits of X rays emitted by neutron stars physicists can set limits on the abundance of such monopoles.

vinced they must be even rarer than the Parker bound indicates. Their reasoning goes as follows. Neutron stars are like gigantic nuclei made up of neutrons, the neutral partner of the proton, and neutron stars have large magnetic fields. Monopoles flying around the galaxy would get trapped by the neutron star's magnetic field and fall into the star. Inside the neutron star they encounter lots of neutrons, which, like protons, they cause to quickly decay into lighter particles. The monopole acts as a catalyst and eats neutrons the way Pac-Man eats video dots. This process releases lots of X rays from the neutron star. Hence the X-ray luminosity of neutron stars as observed by X-ray astronomers provides a limit to the number of monopoles that are flying around the galaxy and getting trapped by neutron stars. X-ray fluxes measured on the old pulsar PSR 1929+10 using the Einstein X-ray satellite limit the number of magnetic monopoles in it to fewer than 1,000 billion (10^{12}). This limit implies an improvement of the previous Parker bound by a factor of 10,000 billion, and if such a bound is at all reasonable, then monopoles are so rare it is completely hopeless ever to expect to find one here on earth. This line of reasoning which brings together the properties of GUT monopoles, proton decay, neutron stars and X rays again exemplifies the powerful constraints that astronomical observations place on the field theories of the microworld.

Physicists draw several conclusions from their experiments, observations and theories of magnetic monopoles. One possibility is that lots of monopoles exist but they are very localized and hidden inside the earth, the stars or the black holes in the center of galaxies. Then the previous bounds on the presence of monopoles, since they refer to the *average* presence of monopoles, do not apply. Another possibility is that the big-bang idea and GUTs are simply wrong. Finally —and this is the option most theoretical physicists currently favor— something got rid of the GUT monopoles shortly after they were created in the very early universe.

If, after the monopoles were made, there was an immense expansion of the universe just prior to the big-bang era, then the monopoles would have been "inflated away." If so, their rarity today would be consistent with all the astronomical bounds. This mechanism accounting for the absence of monopoles supports the "inflationary universe" model—a model I will describe in a subsequent chapter.

But even if monopoles do not exist, the attempt to understand them has already taught physicists a lot. Theoretical physicists, attempting to mathematically master the GUT monopoles, have been led into the forbidding but beautiful mathematics of topology. Rich, complex topological features of field theories are now revealed, and it is hard to imagine that future theories of the microworld will not incorporate these features in some way.

These explorations have revitalized the symbiosis of pure mathematics and physics. Mathematicians and theoretical physicists are talking to each other, the physicists eager to learn the trade secrets of the mathematicians and the mathematicians intrigued by the prospect that their abstractions could contribute to our understanding of the real world. It is a dialogue with a long tradition stretching back to that ancient time when people first grasped that the ratios of simple numbers accorded with the harmonic music of strings. Today, the music of the microcosm is guided by more abstract principles than simple numbers. Yet the abstract music we hear today moves us as the first simple harmonies must have moved those ancient people who first heard the strings sing.

3
Unifying Gravity

The gravitational force is the oldest force known to man and the least understood.

—Peter van Nieuwenhuizen, 1981

Newton discovered the law of universal gravitation that relates the fall of objects on the earth to the motion of the moon and the planets. But his claim to priority for the discovery of this great law was contested. His fellow member of the Royal Society Robert Hooke felt he had preceded Newton—a claim that had some merit, and Newton knew it. Newton, not a timid man, argued against Hooke's claim. The contents of their letters superficially resemble polite correspondence between two English gentlemen, but reading between the lines one finds much insult and bitterness. It was in one such letter to Hooke that Newton made his famous statement "If I have seen farther than other men it is because I have stood on the shoulders of giants." This remark is all the more pointed when one realizes that Hooke was short. Newton not only meant to diminish Hooke's scientific stature but with the same sentence represented his own intellectual lineage as stretching back to the ancients.

Newton's law of gravitation stood as the exemplar of a classical physics law for two centuries. Then on November 6, 1919, before a joint meeting of the Royal Society and the Royal Astronomical Society, observations of the bending of light around the sun were reported, confirming Einstein's theory of gravitation. Newton's law was replaced.

According to Einstein's new theory of general relativity, what we experience as the force of gravity can be mathematically represented by the curvature of a four-dimensional space-time: gravity is geometry. Einstein deduced from his geometrical theory of gravity a deflection of the path of light by the sun double the amount obtained by application of Newton's theory. This larger deflection was what was in fact observed and reported in 1919. Since that time Einstein's general theory of relativity has been repeatedly subjected to stringent experimental tests utilizing, among other modern methods, radar ranging to distant planets and artificial satellites. Again and again it has survived these tests. Most physicists are so impressed with the quantitative success of general relativity in the macroscopic realm that they would be surprised if the theory were to fail dramatically in the near future.

Yet in spite of the experimental success of Einstein's gravity theory in the solar system and the comprehensive picture of the cosmos it implies, from the viewpoint of quantum physics the theory is extremely puzzling. According to quantum-field theory every field has an associated quantum particle, and the quantum of the gravitational field is called the "graviton." A graviton particle, though in theory it exists, interacts so extremely weakly with other matter (far more weakly than neutrinos) that even vast improvements on today's technology would not produce instruments that would detect it.

Nevertheless, theoretical physicists can mathematically calculate the graviton's interactions with other matter, and when they do this, they encounter mathematical infinities in the calculations. Such infinite numbers cropped up before when theorists were calculating the interactions of photons with matter. Amazingly, those photon-interaction infinities could be tamed, "renormalized away," in what appears to be a mathematically consistent procedure. But the infinities encountered in the graviton's interactions could not be "renormalized away"; they were far worse. Quantum gravity interacting with matter is not a renormalizable theory, which simply means that physicists cannot make sense of it.

If the problem with mathematical infinities in quantum gravity theory is insufficiently distressing, there are still deeper conceptual problems. One problem is that the very definition of a quantum particle, which can be precisely formulated in Einstein's special relativity theory, fails or must be dramatically modified if it is to be consistent with general relativity theory. When the quantum theory was combined with special relativity, far-reaching new concepts of the microworld were born. No one knows how these concepts must be modified when quantum theory is combined with general relativity theory. No one yet knows if such a combination is possible.

A simple example illustrates the nature of the difficulty of combining quantum notions with general relativity. The starting point of general relativity is the "equivalence principle"—the principle that a local gravitational field is indistinguishable from an accelerated motion. If you were in a rocket uniformly accelerating in outer space, you would be pressed to the floor just as if a true gravitational field existed in the rocket (as it would if the rocket rested on the surface of a planet).

Einstein recognized in this equivalence principle that the presence of a local gravity field is only an artifact of whether or not an observer is accelerating; that is to say, it depends on the coordinate system with which he chooses to measure his motion. For example, if we choose for the coordinate system the accelerating rocket, then there is a "gravity" field, but in a coordinate system which is not accelerating there is none. But the fundamental mathematical laws of physics should take the same form for all observers independently of whether an observer is accelerating, standing still or moving in any way whatsoever with respect to another observer. Otherwise the fundamental laws would depend upon the arbitrary choice by an observer of a system of measuring coordinates, and that kind of arbitrariness should not be reflected in *fundamental* laws. This principle of "general coordinate invariance" is incorporated into general relativity. In this regard, it goes beyond Einstein's earlier special relativity theory, which required only that the mathematical laws of physics have the same form for observers with uniform motion relative to each other—a special motion at a constant velocity.

According to relativistic quantum-field theory, a constant gravity field creates a radiant bath of quantum particles such as photons at a definite temperature. It would be as if one were inside an oven (fortunately, for the strength of gravity encountered on earth this temperature is very low). But the equivalence principle implies that a gravity field is the same as an acceleration; therefore an accelerating observer sees a bath of quantum particles created by the "gravity" field while one at rest does not. Thus the very notion of creation and destruction of quantum particles is altered. It is unclear what will remain of the concept "quantum particle" in general relativity, yet today that concept is central to physicists' thinking about the microworld.

Theoretical physicists, including Einstein, who have thought deeply about these problems are convinced that combining quantum theory with general relativity theory will require a substantial modification of our basic ideas of physics. Einstein thought that if the quantum theory and general relativity could be combined, this would result in a unified-field theory of all forces. To this day, in spite of enormous effort by the best minds in physics, no one has succeeded in bringing about this combination in a mathematically and physically consistent form. This delay is not so surprising if we remember that it took physicists decades of intellectual struggle to combine quantum theory with the simpler special relativity theory—a merger resulting in the remarkable relativistic quantum-field theories and a new view of the microcosmos. Conceivably, it could take many more decades before physicists achieve a consistent merger of quantum theory and general relativity. Finding a quantum theory of gravity remains the great unsolved puzzle of modern theoretical physics.

Even in the absence of a viable quantum gravity theory, physicists cannot resist speculating about the nature of space-time at distance scales of 10^{-33} centimeters—the Planck scale, at which quantum gravity effects become important—if only to convince themselves that their most cherished concepts break down at these distances. Some physicists maintain that at the Planck scale, space-time develops a foamlike structure. On large-distance scales such as those we experience every day, space-time looks flat and smooth like the ocean surface seen from on high; but up close at the Planck scale it is churning and foaming like the ocean in a storm. If physicists are to describe the microworld at Planck-scale distances, then little remains of the concept of a space-time continuum upon which the description of nature has so far been based. Perhaps some new concepts beyond space and time will be invoked.

Yet in spite of the fact that current relativistic quantum-field theories must fail at the Planck scale, physicists know of nothing that prevents them from describing the microworld for all distance scales larger than the Planck scale. For that reason it may be safe to ignore gravity in thinking about unifying all other forces. Already mathematically consistent theories—GUTs—exist that unify the electro-magnetic weak and strong forces at distance scales before the Planck scale is reached, although they remain to be tested.

Many physicists are convinced that while theories like the GUTs have shed light on the dynamics of the very early universe, not until a totally unified theory exists—one that includes gravity—can they provide an account of the very origin of the universe. For if we imagine going back in time to the very early universe, the tempera-ture and the interaction energy of the quantum particles can increase without limit so that eventually the Planck distance scale is probed. The problem of quantum gravity seems unavoidable if we are to understand the very origin of the universe.

Although physicists are far from achieving the goal of inventing a completely unified field theory that includes gravity, many feel that they have moved closer in the last decade. Their starting point has always been Einstein's theory of general relativity and its associated concepts because that theory so successfully accounts for macroscopic gravitational physics. The problem is how to modify this theory in some way so that one does not lose the successful predictions of large-scale gravity while at the same time resolving the puzzles of short-distance quantum gravity and unifying gravity with the other forces of nature. Two suggested answers to this problem, the "super-gravity theory" and the "Kaluza-Klein theory" of space-time with more than four dimensions, have attracted much recent attention. Whether these ideas are leading physicists down blind alleys or are steps toward the master theory of the universe only time will tell. But they are wild ideas that could be crazy enough to be right.

SUPERSYMMETRY AND SUPERGRAVITY

Every quantum particle can be imagined as a little spinning top, the spin taking on only the discrete values 0, ½, 1, ³⁄₂, 2 and so on in certain units. Spin zero means the top does not spin. Spin ½ means a specific amount of spin, and spin 1 means twice this amount and so on. The pion, a strongly interacting hadron, has spin 0; the proton, neutron, quarks and leptons all have spin ½, while the photon and weak gluons W and Z have spin 1 and the graviton has spin 2. Spin was one of the cornerstones of Wigner's classification system for the quantum particles.

Physicists divide the spinning quantum particles into two classes, the "bosons," which are particles with integer spin—0, 1, 2 and so on—and the "fermions," with one-half-integer spin—½, ³⁄₂ and so on. The reason for this division is that bosons and fermions behave very differently according to the laws of quantum mechanics. Identical bosons, for example, can occupy the same position in space while identical fermions cannot. Identical bosons are "sociable" and prefer to condense in groups, and identical fermions are "antisocial," excluding one another. This "exclusion principle" of fermions is exhibited by electrons, which, as they orbit the nucleus, cannot occupy the same energy state. The exclusiveness of electrons results in a mutual repulsion and explains why atoms do not collapse if squeezed together.

Both bosons and fermions play important roles in the description of nature at the most fundamental level. The standard model has both kinds of particles—the quarks and leptons are spin-½ fermions, while the gluons and Higgs particles are spin-1 and spin-0 bosons. The standard model also possesses important symmetries like the Yang-Mills gauge symmetry which relates the various gluons. But a crucial feature of all such symmetries, which view the various quantum fields as components of a single underlying field, is that the various quantum fields transforming one into another by the symmetry operation must have the same spin. All the Yang-Mills fields, for example, have spin 1, so that such symmetry operations do not mix up fields of different spins.

Recently, to many people's surprise, theoretical physicists discovered another mathematical symmetry that does transform fields with different spin one into another and have given this symmetry the grand name "supersymmetry." Under a supersymmetry transformation, spin-0 boson fields can be transformed into a spin-½ fermi field and vice versa. Just as different gluons could be viewed as different components of a single Yang-Mills field, so according to supersymmetry bosons and fermions of different spin can be viewed as different components of a single "superfield." Under a supersymmetry operation the different components of the superfield, fields of different spin, transform one into another. For the first time the mathematical imagination saw the possibility that all quantum particles, not just those of the same spin, are components of a single master superfield. It is this prospect of unifying particles and fields of different spin, which could include all possible fields, that excites physicists as they explore the mathematical complexities of super-symmetry.

Supersymmetry was discovered independently by several groups of physicists. It was discussed by Y. A. Golfand and E. P. Likhtman at the Lebedev Physical Institute in Moscow and later by D. V. Volkov and V. P. Akulov of the Physical-Technical Institute, Khar-kov. Pierre M. Ramond and John Schwarz of Cal Tech and André Neveu of the Ecole Normale Supérieure also described a boson-fermion symmetry. But not until 1973, when the physicists Julius Wess and Bruno Zumino invented a simple, renormalizable, relativ-istic quantum-field theory that was supersymmetric, did supersym-metry attract the attention of many physicists. Since that time industrious theorists have constructed many other supersymmetric-field theories.

The Wess-Zumino supersymmetric-field theory was not intended to be a realistic mathematical model of existing quantum particles. Instead it provided a kind of conceptual laboratory in which mathe-matical physicists could explore the implications of supersymmetry before moving on to the construction of more complicated models which they hoped might be experimentally relevant. The fundamen-tal mathematical object in this model was a single superfield with components corresponding to spin-0 fields and spin-½ fields. Exact supersymmetry implied that the spin-0 and spin-½ quantum particles must have equal mass (this also is true in more complicated super-symmetric models). But such equal-mass particles of different spin have never been observed in nature, and that is why the simple model is not experimentally relevant. Supersymmetry, if it is to lead to a description of the particles observed in nature, must therefore be broken. Then the masses of particles of different spin related by supersymmetry need not be equal. Yet even if physicists make math-ematical models in which supersymmetry is broken, they have not succeeded in relating any of the presently detected particles of differ-ent spin to one another by supersymmetry.

Physicists who have studied supersymmetric theories are con-vinced that ordinary quarks, leptons and gluons, although they have different spins, are *not* related to each other by a supersymmetry operation. If broken supersymmetry is to be manifested in nature, then the "superpartners" of quarks, leptons and gluons—the quanta that are indeed related to them by supersymmetry—must be alto-gether new quantum particles, none of them so far detected and most of them very massive. Theoretical physicists are having fun predicting their properties.

If nature is governed by supersymmetry, then the microworld is organized by a kind of "supermirror." On one side of the supermirror are the ordinary particles like leptons, quarks and gluons; on the other side of the supermirror, each of these particles has a superpartner image—new particles dubbed "leptinos," "quarkinos" and "gluinos." While the leptons and quarks are spin-½ fermions, their superpartner images are bosons. The superpartner of the spin-1 pho-

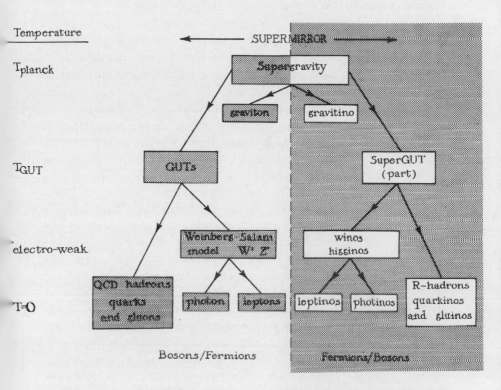

The Supermirror. If supersymmetry exists, then the ordinary quantum particles —quarks, leptons and gluons—have superpartners—quarkinos, leptinos and gluinos, which could be very massive. Perhaps they can be detected when new accelerators are built. This diagram illustrates the unification of forces as the temperature of the universe rises. At the Planck temperature, all the forces are unified under the aegis of a supergravity theory. At lower temperatures, the supersymmetry is broken, giving rise to the differences between ordinary particles and their superpartners.

ton is the spin-½ "photino," and so on. The enterprise of naming these imagined particles has something in common with the naming of imagined beasts like manticores and unicorns by medieval zoologists.

Perhaps when the new supercolliders like the desertron turn on, the energy will be sufficient to create the new superparticles and these imagined "beasts" will become real. Unfortunately, no one can confidently calculate the expected mass of the superpartners, and hence the energy threshold for producing these exotic new particles is unknown. Conceivably, the mass of most superpartners is so high that even supercolliders cannot create them; perhaps the only time that sufficient energy was available was in the early universe.

A few superparticles like the photino and the "gravitino," the superpartner of the graviton, might have quite low masses. If so, then the reason they have not been directly detected is not that they are so massive but that they interact so extremely weakly with ordinary matter. These particles may play a role in the evolution of the universe. Some physicists speculate that photinos or gravitinos liberated from their interactions with the rest of matter in the early stages of the big bang could be the dark matter in the universe today; also, by gravitational clustering they could have played an important role in galaxy formation. The structure of the universe itself becomes a proving ground for the wild ideas of supersymmetry.

Supersymmetry as manifested in the prototype Wess-Zumino model is "global" supersymmetry, meaning that the symmetry operation is the same over all of space. Zumino in his papers and lectures emphasized that just as the older internal global symmetries could be generalized to local Yang-Mills–type symmetries, so too global supersymmetry ought to have a local gauge-field generalization. Furthermore, it was apparent from the mathematical properties of supersymmetry that such a generalization would bring in the gravitational field as a gauge field associated with local supersymmetry. The first mathematical theory of such a local version of supersymmetry, dubbed "supergravity," was discovered in 1976 by Sergio Ferrara of Frascati Laboratories, near Rome, and by Daniel Z. Freedman and Peter van Nieuwenhuizen of the State University of New York at Stony Brook. Shortly thereafter, a simpler alternative derivation of supergravity theory was obtained by Bruno Zumino of CERN and Stanley Deser of Brandeis University.

Supergravity turned out to be an imaginative extension of Einstein's theory of gravity, making it a supersymmetric theory. Remarkably, general relativity was generalized. Einstein's original theory could be viewed as describing the graviton, the hypothetical quantum of gravity, as a spin-2 boson. In the supergravity extension of Einstein's theory, the graviton acquires a superpartner, the gravitino, a spin-$\frac{3}{2}$ fermion, and under local supersymmetric transformations these two particles transform one into the other.

When theorists did quantum calculations using supergravity theory, they discovered to their surprise that the infinities that plagued the earlier gravity theory with only the graviton were now being cancelled by equal and opposite infinities produced by the gravitino. Such cancellations were not fortuitous but a deeper consequence of the presence of supersymmetry. Although it is not yet known if the supergravity theory is completely renormalizable, this "softening of the infinities" seems to be a step toward a viable theory of quantum gravity.

Simple supergravity theory includes only the graviton and the gravitino, and this hardly corresponds to the real world with its many particles. Are there other mathematical supergravity theories? Physicists, in their search for mathematical theories of the quantum particles, are always looking for powerful general principles that might limit their search. They have shown that the principle of local supersymmetry is so restrictive that only eight possible supergravity theories exist. (This is similar in spirit to showing that only five regular solids exist in three dimensions). These eight supergravity theories are labeled by an integer $N = 1, 2 \ldots 8$, where $N = 1$ supergravity is the simplest, with just the spin-2 graviton and the spin-$\frac{3}{2}$ gravitino fields. The higher N supergravity theories require fields with spin 0, $\frac{1}{2}$ and 1 as well. $N = 8$ supergravity, the most complex, has a total of 163 fields all mutually related by supersymmetry. It is tempting to identify some of those 163 fields with the known quarks, leptons and gluons of the standard model, but unfortunately that simple identification fails. Most people who have worked on supergravity feel that some crucial idea is still missing; without it the theories simply don't describe the real world.

In spite of the unsolved problem of making supergravity theory realistic, theoretical understanding has advanced. Each of these theories brings in gravity, so that supergravity is potentially a completely unified field theory. All fields, including spin-0, -$\frac{1}{2}$ and -$\frac{3}{2}$ fields, are now a consequence of a local supersymmetry, whereas previously, only spin-1 Yang-Mills fields could be deduced as a consequence of symmetry.

Physicists who have worked extensively on supergravity testify to its underlying conceptual power and mathematical complexity, features it shares with its progenitor, the general relativity theory. Perhaps by postulating the existence of a single master supersymmetry, physicists can account for the whole universe. It is yet too soon for them to give up on that dream. Perhaps some supersymmetric theory is the "Holy Grail" of the physicist's quest.

THE FIFTH DIMENSION AND BEYOND

One feature of our physical world is so obvious that most people are not even puzzled by it—the fact that space is three-dimensional. In Einstein's special relativity theory, space and time become intimately intertwined, so much so that Hermann Minkowski was able to show that time in this theory might be viewed as a fourth dimension (although it is not a spatial dimension). No one has the slightest idea why the world we live in has one time and three space dimensions and not, for example, eleven dimensions. Of course the world would be extremely different if we altered its dimensionality. Maybe higher dimensions are lethal to life and we should be thankful for our modest allotment of four.

The fact, confirmed by experience, that the world we live in has three plus one space-time dimensions is simply written into the laws of physics as they now stand. Some physicists are dissatisfied with this and feel that the dimensionality of our world should be logically deduced from a master theory of the universe and not be a starting postulate. Today these physicists cannot yet calculate the observed number of space-time dimensions from first principles. But, remarkably, they are developing a conceptual framework in which such a calculation might someday make sense. That conceptual framework, known as the "Kaluza-Klein theory," grew out of yet another generalization of Einstein's four-dimensional general relativity, this time to higher-dimensional spaces. Before describing the remarkable Kaluza-Klein theory, I will make a short digression by describing what is meant by "big" and "small" dimensions.

The three spatial dimensions we observe are "big" dimensions—

we can walk around in them. If there exist additional dimensions, then they must not be like the "big three"; if they were, we could walk around in them too, and this, clearly, conflicts with experience. The extra dimensions that physicists are contemplating are "small" dimensions, so small they cannot be seen and hence do not directly influence our three-dimensional perspective of the world. What are "small" dimensions?

To visualize "small" dimensions, imagine a world that has only one "big" dimension. The space of this one-dimensional world is represented by an infinitely long line. Next imagine that this line lies on the surface of a cylinder, so that the complete space is now represented by the two-dimensional surface of the cylinder. The second, "extra" dimension corresponds to going around the cylinder. If you do this, you come back to the point you started from—the extra dimension is a circle, not a line.

Spaces that curl up on themselves like the one-dimensional space

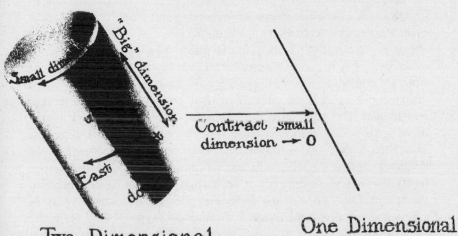

Two Dimensional Cylinder Space

One Dimensional Line

The surface of a cylinder is a two-dimensional space. The "big" dimension is the line and the "small" dimension is the circle. If the small dimension is shrunk to zero radius, then there remains just a line, a one-dimensional space. A similar idea may apply to our world, which may have more than four dimensions. The other, higher dimensions could be "small" dimensions which we do not observe directly.

of a circular line or the two-dimensional surface of a sphere are called "compact spaces" by mathematicians. A cylinder can be thought of as a two-dimensional space, one dimension of which is compact (the circle) and the other of which is noncompact (the line). We can imagine that the radius of the circle is so small it shrinks to zero; then we are back to just a one-dimensional space—the infinitely long line. Clearly, by making the circle very small we can approximate the one-dimensional space of the line as closely as we like. The circle is the "small" extra dimension and the line is the observed "big" dimension.

The possibility that there exist extra "small" dimensions beyond the "big four" of space-time—dimensions so small that they are not in conflict with experience—was discovered in the context of Einstein's general relativity by Theodor Kaluza in 1919 (the paper appeared in 1921). Kaluza, a mathematician and linguist, examined Einstein's equations generalized to a five-dimensional space-time in which the "extra" fifth dimension was compact—just a small circle. He imagined that at each point in ordinary four-dimensional space-time there is a little circle, just as on every point along the line on the cylinder we considered there is a little circle.

Just as we can move from point to point in ordinary space, we can imagine a particle moving around the little circle in the fifth dimension. Of course, it doesn't move very far (and not at all in the "big" dimensions), because the circle is so small and all it does is go around and around. But still, what does the possibility of this extra movement mean? Kaluza showed that this extra freedom of movement associated with a circle symmetry at each point in space-time could be interpreted as the simple gauge symmetry of the electromagnetic field. This interpretation is not so surprising to us from a modern viewpoint if we realize that a symmetry (like the little-circle symmetry) automatically implies the existence of a gauge field (like the electromagnetic field). The five-dimensional Kaluza theory thus not only described the curvature of the big four-dimensional space-time in terms of the usual Einstein gravitational equations, but also physically unified gravity with Maxwell's electromagnetic gauge field using the strange idea of a fifth, circular dimension. Kaluza's accomplishment impressed Einstein, who wrote to him, "The idea of achieving [a unified theory] by means of a five-dimensional cylinder world never dawned on me. . . . At first glance I like your idea enormously." Einstein himself began to work on the idea.

Kaluza demonstrated the unification of gravity and electromagnetism by means of his compact fifth dimension only by making several restrictive assumptions in solving Einstein's equations. In 1926, Oskar Klein significantly advanced this theory by showing that these restrictive assumptions were completely unnecessary. Furthermore, Klein calculated the radius of the little circle in the fifth dimension in terms of the known quantities, the Planck distance scale and the electronic charge, and found this radius to be about 10^{-30} centimeters —an extremely small radius ensuring that the fifth dimension is safely out of sight. But in spite of its small size, the freedom that fields have in moving around that tiny circle is always present at every point in ordinary space, and that freedom is all that is needed to guarantee the existence of the electromagnetic field.

After the 1930s the Kaluza-Klein idea fell out of favor, and for many years it lay dormant. But recently, as physicists searched out every possible avenue for the unification of gravity with other forces, it has again sprung to prominence. Today, in contrast with the 1920s, physicists are challenged to do more than unify gravity with just electromagnetism—they want to unify gravity with the weak and strong interactions as well. This requires even more dimensions, beyond the fifth.

Theoretical physicists have generalized the original five-dimensional theory to an arbitrary number of higher dimensions. All the higher dimensions are compact; they are curled up into a tiny multidimensional space that exists at each point of ordinary space and hence is unobservable. But the freedom of moving around these compact tiny spaces with symmetries more general than the simple symmetry of a circle corresponds exactly to the freedom of performing Yang-Mills gauge transformations. Remarkably, the local gauge symmetries are precisely the symmetries of the compact higher-dimensional space. Because of this mathematical fact, all the gauge theories of Yang-Mills fields can be interpreted purely geometrically in terms of such compact higher-dimensional spaces.

Unfortunately, the Kaluza-Klein theory is extremely restrictive, so restrictive that no one has yet been able to find a realistic Kaluza-Klein theory which yields the standard model. Physicists, while they welcome such restrictive principles to limit their choices in the search for the correct theory, are frustrated by the realization that such extreme restrictions have so far led only to theories that fail to describe the observed quantum world. Nonetheless, the search goes on.

A remarkable discovery was made in 1978 by Eugene Cremmer and Bernard Julia, two French mathematical physicists, when they combined the Kaluza-Klein idea with supergravity theory. Recall that there are eight distinct supergravity theories, with $N = 1$ supergravity the simplest, with just the graviton and gravitino fields, and $N = 8$ the most complicated, with 163 different fields. Cremmer and Julia noticed that if one examines $N = 1$ supergravity in an eleven-dimensional space (instead of a four-) and assumes that seven of these eleven dimensions are compact, à la Kaluza-Klein, and the remaining four are the "big" space-time dimensions, then the resulting theory in these four dimensions is $N = 8$ supergravity. A simple $N = 1$ supergravity theory in eleven dimensions becomes the complicated $N = 8$ supergravity theory in four dimensions. This result encourages workers who hope that the complex field theories they require to describe the real four-dimensional world are the offspring of much simpler theories when viewed in higher dimensions. Some physicists hope that all they need to do is find the appropriate application of the Kaluza-Klein idea and the master theory of the universe will appear.

In spite of the aesthetic appeal of the central ideas, there are major mathematical obstacles that must be overcome if the multidimensional unification idea is going to work. For one thing, no one knows the deep reason why some dimensions are compact and small and others—the four that we see—are large. Even in the Kaluza-Klein theory it is simply assumed that four dimensions are big while the others are compact—an assumption which physicists hope may someday be removed. Conceivably the idea of broken symmetry— here the broken symmetry of a high-dimensional space—plays a role in liberating them from this assumption. Perhaps the real world with its four big dimensions corresponds to the broken but stable solution to equations describing the symmetries of a multidimensional geometry. Such hints, while intriguing, have not yet solved the outstanding puzzle of the observed dimensionality of space-time.

There are other problems. Klein in his pioneer work calculated the radius of the fifth dimension in terms of the Planck length and the electronic charge, a measure of the strength of the electromagnetic interaction. If one knew the value of the radius of the fifth dimension, then by turning the calculation around one could calculate the electronic charge. Recently physicists have estimated the radii of the other dimensions and used this to calculate the charges which measure the strength of other forces. But these calculated charges are far too large to have anything to do with the observed strength of the forces. This suggests that these multidimensional theories are not very realistic.

Yet another puzzle is that the spin-½ fields we observe in nature are difficult if not impossible to accommodate in a multidimensional compact space. Some crucial ingredients are lacking before the Kaluza-Klein idea can be made to work—if it can work at all.

Such puzzles goad the imagination of theoretical physicists today. The idea that the many gauge symmetries which play such a crucial role in an understanding of the forces of nature are simply a manifestation of the symmetry of a higher-dimensional space has such appeal that work on this marvelous idea will continue until it is finally shown to be irreconcilable with experience or until a much better idea comes along. The hope of accomplishing a geometrical unification of gravity with the other forces of nature through a profound extension of Einstein's theory of general relativity to many dimensions will not be easily forsaken. Perhaps one day persistent physicists may even solve the puzzle of why our world has three space dimensions and one time dimension. In the arena of such unleashed wild ideas, even the solution of that deep puzzle does not seem too much to expect.

Now let us see what these wild ideas, if they are right, imply about the origin of our universe. Let us press on to the world before the first nanosecond!

4
Before the Big Bang: The Inflationary Universe

I have often heard it said that there is no such thing as a free lunch. It now appears possible that the universe itself is a free lunch.
—Alan Guth, 1982

The big-bang model of the early universe covers a period from about the first nanosecond to the first 300,000 years and correctly predicts the relative abundance of the light elements and the temperature of the microwave background radiation. These successes promoted a deeper investigation into the underlying assumptions of the big-bang model. And when physicists did this, they confronted some puzzles about the very nature of the whole universe.

If we continue to use the big-bang model as a description of the universe before the first nanosecond, then the supercomputer on which we imagined programming the model has an extremely simple output: a homogeneous gas of all the quarks, leptons and gluons getting hotter and hotter as the universe contracts to a singularity. Even if we ignore the problem of the bizarre singularity at the beginning of time, other problems arise in the big-bang model if we want to account for some of the most dramatic features of the observed universe—its large-scale isotropy (the fact that it appears much the same in all directions) and its spatial flatness. We have simply put these observed features into the standard model as beginning assumptions when we assumed an isotropic FRW geometry for space-time with a value of Ω near one—equivalent to assuming approximate spatial flatness. It is worth reminding ourselves of the observational evidence supporting these assumptions, for, as we will see, they are very puzzling assumptions.

On the largest distance scales the universe is remarkably homogeneous and isotropic, as revealed by the observed even distribution of galaxies and quasars in the sky (rather than their being gathered together in a few regions). The best evidence for the isotropy of the universe is the microwave background radiation, which may be isotropic to better than 1 part in 10,000. Since this radiation is a fossil of the big bang, we can conclude that the early universe during the big bang was also extremely isotropic. If we want to account for this remarkable isotropy from the beginning of the universe to the present time, then we are left with a puzzle which is easily illustrated.

Suppose the universe began 12 billion years ago. Looking out into the sky one night, we spot a group of quasars 7 billion light-years away. If we look in exactly the opposite direction in the sky we spot another group of quasars, resembling the first group, which is also 7 billion light-years away. Reflecting on these facts, we see there is a puzzle. The two groups of quasars are a total of 14 billion light-years away from each other, and since the universe is only 12 billion years old, light traveling from one group of quasars cannot possibly have reached the other group. Because light travels at the maximum speed, we conclude that the two clusters of quasars are "causally disconnected"—they cannot ever have communicated or influenced each other at any time in the past. If this is so, then why do they look so similar in their appearance and state of evolution, just as if they possessed a common origin and were subjected to the identical physical processes long ago?

To appreciate how peculiar this is, imagine being on a movie set—a large landscape—on which a battle scene is about to be filmed. All the actors and extras have been precisely instructed to coordinate their actions so that the result will be the appearance of a real battle. This requires careful planning and communication between the different groups of people and the director. Imagine what would happen if such planning were not made or were impossible. One would expect chaos, with people and groups showing up at the wrong times and places. We would be quite surprised if the result was a perfect movie scene.

We ought to be similarly surprised that different groups of galaxies and quasars in the universe's "landscape" which have never had the opportunity to communicate with each other look so much alike. The isotropy of the universe is indeed amazing. For example, the temperature of the microwave background radiation is nearly the same in every direction of the sky, in spite of the fact that this radiation originates from billions upon billions of regions of the universe that were causally disconnected in the big bang. How were all those distinct regions "precisely instructed" to have the same temperature today if they cannot ever have communicated? Likewise, what "told" the galaxies and quasars to be distributed so uniformly? The observed isotropy of the universe is thus a puzzle of causality. From the viewpoint of the big-bang model, isotropy is a complete accident.

Of course, it is possible to set up the initial conditions in the mathematical model of the early universe just right so that when causally disconnected parts of the universe first contact each other the result will be a homogeneous and isotropic universe. This is what is assumed in the standard big-bang model. But such "fine tuning" of the initial conditions of the universe to produce the observed result does not provide an explanation of the uniformity. Really it is we—the model builders—who artificially put in the uniformity to begin with.

The observed uniformity of the universe is but one puzzle posed by the standard big-bang model. Yet another puzzle is the observed near flatness of the large-scale space of the entire universe. At first we may think that this flatness is simply a natural assumption. But according to the general theory of relativity, the space of the entire universe can curve and its curvature is related to the cosmic parameter Ω, the ratio of the average density of matter to a known critical density of matter. If Ω is equal to 1, then space is flat; if Ω is greater than 1, then space has a convex, positive curvature; and if Ω is less than 1, space has a concave, negative curvature.

Astronomers estimating the average density of matter find a range of values for Ω anywhere between $1/10$ and 2, so that the observations are not sufficiently refined to make it possible to determine which one of the three curvature possibilities may actually be realized. Yet the remarkable fact, often overlooked, is that Ω is found to be so close to unity—even if space is curved, it is extremely close to the flat space for which $\Omega = 1$. Why is this so? In theory the parameter Ω could have any value. It could be as big as 10,000, in which case the universe would have collapsed long ago, or it could be as small as $1/10,000$, in which case matter would be so dilute that galaxies could not exist.

This flatness puzzle is compounded by the realization that only by "fine tuning" of the initial conditions of the universe to one part in billions of billions does the value of Ω come out to lie in the range of values estimated by astronomers today—values near to unity. This would be like "fine-tuning" the initial firing of a bullet from a gun so that the bullet eventually comes gently to rest in a child's hand. Such "fine tuning," while possible, hardly provides a satisfactory explanation for the observed flatness of the universe.

Yet a third feature of our contemporary universe is puzzling from the standpoint of the big-bang model: the absence of magnetic monopoles, topological solitons, twisted lumps of field energy as described by the unified-gauge-field theories—the GUTs. According to the GUTs, the matter in the very early universe when it was only 10^{-35} seconds old consisted of a gas of quarks, leptons and gluons at very high temperature, interacting symmetrically, without distinction between the various forces. As the universe expands and cools the temperature drops, the symmetry is broken and the various interactions become distinct. At some critical temperature the electromagnetic interaction becomes distinct because the unifying symmetry is broken. When that happens, magnetic monopoles can be created in the early universe. It is not difficult to understand why.

Recall our illustration of spontaneous symmetry breaking—a group of people sitting down at a round dinner table and required to pick a salad plate to either their right or their left, thus breaking the right–left symmetry. If one person, out of the many seated, picks the salad plate to his right, and another person, seated elsewhere, picks the dish to the left, then it follows that at least one dish will be claimed by two people and one dish by none. This broken symmetry configuration contains at least two "topological twists," one located at the dish claimed by two people (the soliton) and the other at the dish claimed by none (the antisoliton). Interestingly, if we ask one of the two people claiming the same salad plate to alter his choice, and then his new neighbor does likewise (and so on), the location of the dish claimed by two people (the soliton) will move around the table until it meets the antisoliton, annihilating it. Then everyone at the table has one salad plate. The right–left symmetry is still broken, but it is now broken in the same direction around the entire table.

In order to develop the analogy of symmetry breaking at the dinner table with symmetry breaking in the early universe, we require an analogue of high temperature for the dinner table. The effect of

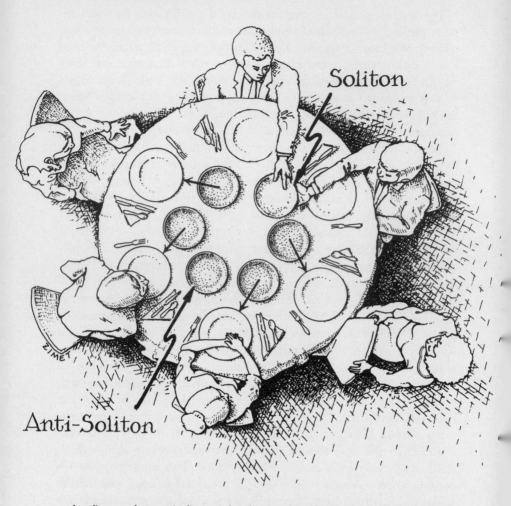

A soliton and an antisoliton at the dinner table. If one of the people claiming the same plate instead choses the adjacent plate, then the soliton begins to move around the table until it meets the antisoliton and annihilates it.

high temperature would be analogous to the people at the table being very agitated and confused so that they are continually changing their choice of salad plate from right to left and back again. Then, on the average, neither right nor left choice is preferred and the situation is symmetrical, an example of symmetry restoration at high temperature. As the temperature drops, this corresponds to the people's becoming less agitated and settling upon which salad they want. The symmetry is now broken. But when it breaks some people may pick the dish to the right, others to the left, and this, we see, produces the topological twists—the solitons. Likewise, in the early universe, as the temperature falls symmetries are broken. But if they break differently in different regions of the universe, then they can form topological solitons, the magnetic monopoles. How many such monopoles and antimonopoles would be produced?

Suppose the round dinner table is so large that it cannot fit into one room but intersects many rooms, so that there are walls between groups of people seated at the same table. All the people in any one room, as they become less agitated, will each pick a salad plate, and since they can all communicate with each other, all agree to pick a dish on the same side. But since they cannot communicate between rooms, the people in adjacent rooms have a fifty-fifty chance of picking a different dish. If the walls are now removed, we would estimate that there are one-half as many twists as there are rooms.

The very early universe imitates this illustration. The different rooms correspond to causally disconnected regions of the universe which, as the universe evolves in time, come into causal contact. Since broken symmetry fields can be aligned differently in different regions, the way the choice of salad plates can be different in separate rooms, magnetic monopoles are produced when the regions contact each other. The universe is expanding sufficiently rapidly so that the monopoles and antimonopoles are moving rapidly apart and do not mutually annihilate very much. Physicists calculating the number of monopoles produced in this way estimate that the mass density of the universe today should be dominated by such magnetic monopoles. This is an absurd result, especially in view of the fact that no monopoles have ever been detected on earth, and there are stringent bounds on their abundance in the universe. So the big-bang model, along with some ideas about broken field symmetry, poses yet a third puzzle: what happened to all the magnetic monopoles?

All three puzzles—isotropy, flatness and monopoles—are solved if one assumes the existence of a new pre–big-bang epoch of cosmic evolution called the inflationary universe. The signal feature of this rather short but important epoch is that the space of the universe undergoes an immense expansion, far greater than the much slower subsequent expansion during the big-bang epoch. The lifetime of the inflationary universe epoch lasts from perhaps the first 10^{-35} seconds to 10^{-33} seconds, well before the first nanosecond. It's hard to imagine how anything lasting for such a short time so very early in the thermal history of the universe could solve the puzzles of isotropy, flatness and the absence of monopoles. Yet, remarkably, the existence of a very early inflationary epoch does just that.

The story of the theoretical discovery of the inflationary universe begins in the spring of 1979. Alan Guth and Henry Tye, two young theoretical physicists at Cornell University, set out to mathematically study the puzzle posed by the cosmological absence of magnetic monopoles. Guth, like many particle physicists at that time, was skeptical about using cosmology to constrain theories of the quantum particles, but Tye, already a convert to this idea, persuaded him that this approach had merit. By September, Guth and Tye, now at the theory division of the Stanford Linear Accelerator Center, wrote a paper suggesting that if the early universe had a new "supercooling" pre–big-bang phase during which the temperature dropped precipitously, then the magnetic-monopole-abundance problem could be solved. Inflation was not yet part of the picture.

Sometime in the fall of 1979, Guth first asked himself the key question How would such a supercooling phase affect the evolution of the universe? He found that the answer was that before the hot big bang the universe must have had an "inflationary" phase, a period during which space underwent an exponentially rapid expansion in scale size. In the usual big-bang model, the space in the universe also expands but at a much slower rate. In Guth's mind the supercooled universe and inflation were now logically linked. But assuming such a strange inflationary phase seemed to him a farfetched idea at the time, and it explained nothing except the absence of monopoles.

Guth, fortunately, remembered a talk given during a visit to Cornell by Princeton physicist Robert Dicke. Dicke had spoken about work that he and P. James E. Peebles had done which emphasized that the observed flatness of the universe was without explanation in the big-bang model. In November 1979, Guth realized that his new idea of inflation solved this vexing flatness puzzle. Simply assuming the existence of an inflationary epoch predicted a value of $\Omega = 1$—a flat space—and made "fine tuning" unnecessary. This amazing result made the hypothesis of inflation more attractive. But at that time Guth had not even heard about the isotropy-causality puzzle.

In December 1979 or early January 1980, Guth and other theorists at the Stanford Center were discussing a paper by Anthony Zee which emphasized the causality puzzle. Guth almost immediately realized that his inflationary scenario solved that cosmological puzzle as well and that he was on to something.

He then worked out the important details by making an explicit cosmological model based on the SU(5) GUT, a model which implied the existence of a pre–big-bang inflationary epoch. The paper was finished by the end of the summer of 1980 and published in 1981. The idea of the inflationary universe was born, the first really fresh idea in cosmology for many decades. Curiously, Guth had the answer—the inflationary-universe epoch—before he knew of the questions it answered.

Reflecting on this sequence of events some years later, Guth finds it amazing that no one had preempted his theoretical discovery, because all the pieces required to solve the puzzles were out in the open. Wolfgang Pauli, a theorist of a previous generation, complaining about his lack of creativity in his later years, said, "I know too much." Guth, a novice in cosmology, did not "know too much" and in the course of examining the relation between cosmology and quantum-particle physics found a new way to view the very early universe.

In order to visualize the inflationary universe we will develop a simple image. Instead of a curved three-dimensional space of the real universe (which is hard to visualize), imagine the space of the universe as a one-dimensional line, either an infinite line for an "open" universe or a circle for a "closed" universe. For definiteness let us suppose the universe is closed, so its space is a circle. Then the expansion of the universe is represented by the radius of the circle enlarging and the circle stretching like a perfectly elastic rubber band. This is our "rubber-band" model of the universe.

Further, let us imagine that light waves can propagate along this one-dimensional space like small elastic vibrations on a rubber band. The important physical feature of a light wave is that it is the fastest way to send an energetic signal between two points in empty space. Therefore light rays establish whether an event can causally influence a second event—a light ray resulting from the first event must be able to reach the second.

Next imagine that the circle is expanding. The radius of the circle can expand as fast as it wants; it is not limited by the speed of light because no energy is being transported by such an expansion. For example, in the standard model the radius of the universe grows faster than the speed of light; in fact, it grows as a power of the elapsed time.

If a physical event takes place in such an expanding space at an instant of time, then we can imagine that light rays are emitted by that event in all spatial directions. If a second event takes place somewhere else in space before any light ray from the first has reached it, then the second event is "causally disconnected" from the first—there is no way that the first event could have influenced the second. Eventually, in the future, the light rays from the two events might reach each other's location, and they then contact each other for the first time.

This is what happens in the standard model. Many regions of the sky that we now see from our galaxy correspond to events that were causally disconnected from us (and each other) in the early universe. Imagine that in the very early universe, when its radius was very small, there were two tiny fluctuations which eventually grew into galaxies. These two fluctuation events we assume are precisely correlated, so that at later times when they came into causal contact the two galaxies would appear the same. This is "fine tuning" of the initial universe to fit the later facts. But suppose we had set up the initial conditions differently so that one event was some little fluctuation but the other was nothing at all. We can certainly do this, since the two events are not causally connected in any way. Then such an initial condition might result in a universe today for which some large region of the sky had no galaxies at all—an extremely anisotropic universe. The fact that in the standard model the currently observed universe consists of countless billions of regions that were causally disconnected in the early stages of the big bang and therefore independent of each other, and yet are very much alike today and hence apparently correlated, is the isotropy-causality puzzle.

In the context of the simple rubber-band model we will now see how inflation solves not only this vexing causality problem but the flatness and monopole-abundance problems as well. Again, the basic idea of the inflationary universe is that the space of the universe undergoes an immense expansion at a very early pre–big-bang stage. During this brief inflationary period lasting from perhaps the first 10^{-35} seconds to about 10^{-33} seconds, the growth of the radius of the universe is exponential in time as contrasted with the much slower growth rate during the big-bang expansion. To get an idea of the amount of inflation involved, imagine a rubber-band universe the size of a finger ring expanding during the inflationary period to a circle the size of the current observed universe—a factor of 10^{30}. Even that immense inflation is very small compared with the minimum factor of 10^{50} required for the inflation to solve the cosmological puzzles. After this immense exponential inflation stops, the universe continues expanding but at the much slower big-bang rate.

It is easy to see that such inflation solves the causality problem, because any small causally connected region of space at the beginning of the inflation now gets stretched into a huge region of space by the end of the inflation. All the events in that immense region of space are now correlated because they originated from a single extremely small causally connected region.

According to this scenario, the entire currently visible universe, and all of the universe we will see for unimaginable time in the future, originates from a single causally connected spatial region before the inflation began. After the inflation is over, new events and processes can occur in the immense causally correlated region of space—for example, the formation and evolution of galaxies—which are *not* directly causally connected. But all these subsequent events were set in motion by even earlier events before the inflation which *were* causally correlated. These events got their "marching orders" before inflation; although they may subsequently have lost causal contact, they knew what to do.

So from the standpoint of the inflationary universe it should come as no surprise that as we look out into the distant reaches of space we see galaxies that do not appear to be so different from our own galaxy. Our proper greeting as we see these galaxies, however, should not be "hello" but "hello again." For long ago—before inflation—we were all together in the form of primordial density fluctuations.

Inflation also solves the flatness problem, the puzzle of why Ω is so close to unity. The enormous region of space we observe today corresponds to only a very tiny segment of a huge inflated circle. This space, of course, looks like a line, a flat space, and not a circle, just as the earth appears flat if viewed over a small region. A definite prediction of the inflationary-universe scenario is therefore that $\Omega = 1$— the observed universe must be precisely flat today irrespective of its curvature prior to inflation.

The definite prediction of $\Omega = 1$ could be a problem for the inflationary picture because current observations favor a value of Ω of about one-tenth. Furthermore, if one simply assumes that Ω is as big as one, then the calculations of the relative abundances of the elements created during the big bang (which are sensitive to the value of Ω) do not seem to agree with the observed abundances.

In spite of these potential conflicts between observations and the inflationary scenario, most cosmologists think that eventually the prediction $\Omega = 1$ will be borne out. At a 1984 San Francisco meeting of cosmologists and particle physicists, a vote was taken on what these scientists thought the eventual value of Ω would be. In spite of the evidence, the overwhelming consensus was that $\Omega = 1$. While this hardly represents a scientific method of determining Ω, this vote does reflect how deeply the idea of inflation has taken hold.

Inflation also solves the third puzzle—the monopole-abundance problem. Recall that monopoles were produced as topological twists when causally disconnected regions of space (in which symmetry-breaking fields had different orientations) were in contact with each other. But if the entire observed universe today originates from a single causally connected region rather than many such regions, then no monopoles should exist in the present universe. Any monopoles produced in the preinflationary phase of the universe were simply "inflated away"—they became so dilute that they effectively do not exist.

Here we see a primary property of inflation: it dilutes the universe. Any specific quantum particles produced before or during inflation will be "inflated away"—the house of the universe is swept clean by inflation.

The inflationary universe thus solves the three cosmological puzzles posed by the big-bang model. But what might have caused this remarkable behavior of the very early universe—this immense inflation? And why did it stop? I have said nothing about that.

The dynamical behavior of the geometry of the universe is governed by Einstein's equations which relate the curvature of space to the properties of matter in the space. In the big-bang model the matter was represented by a gas of all the quantum particles, and the properties of such a gas according to the standard model do not imply an inflationary phase.

In the simplest inflationary-universe models, like the one first considered by Guth, the matter in the universe is represented by a single spinless field which is uniform over space but whose magnitude

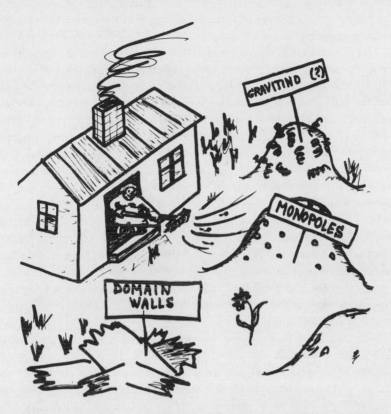

A drawing by A. Linde, a physicist who has helped to develop the idea of the inflationary universe. A desirable feature of the inflation is that the house of the universe is swept clean of unwanted (because they are unobserved) objects like magnetic monopoles, gravitinos (the superpartner of the graviton, the quantum of gravity) and "domain walls," which can form between regions of space with topologically distinct field configurations.

changes in time. Such a spinless field, which can be a symmetry-breaking Higgs field, with all the right properties, is present in many of the unified-field theories. So assuming its existence, although a "wild idea," is not extremely farfetched. The presence of this field produces a uniform mass-energy density and pressure throughout the universe (in our rubber-band model, along the string), and if we can but determine that density and pressure the Einstein equations will specify the curvature of space. Because the magnitude of the field changes in time, the mass-energy density and pressure in the space of the universe also change in time, implying that the curvature changes in time as well.

The mathematical equations that describe how the uniform spinless field changes in time turn out to be identical to those that describe a ball rolling down a hill, so let us draw an analogy between the field and the ball. Because this analogy is mathematically precise, theoretical physicists themselves discuss the dynamics of inflation in terms of the ball and the hill, knowing that they can translate these concepts back into the language of fields. In this analogy, the horizontal position of the ball along the hill is equal to the magnitude of the spinless field, and the vertical height of the ball above the bottom of the hill —its potential energy V—is equal to the mass-energy density of the field. The kinetic energy of motion of the ball as it rolls down the hill is equal to the sum of the mass-energy density and the pressure (measured in certain units) produced by the spinless field in the space of the universe. Hence all that need be done to calculate the mass-energy density and pressure in the universe as it changes in time is to determine how a ball rolls down a hill of specified shape—a simple problem in classical physics.

Next we have to say how the pressure and mass-energy density affect the change of the curvature of the universe. In our simple rubber-band model, that amounts to determining the change of the radius of the circle in time. If the universe is expanding, then the mass-energy in it will slow the expansion in the same way that the mass of the earth, by the influence of its gravity, slows the motion of a stone thrown upward.

The effect of the pressure is more complicated. One might think that a positive pressure in the universe would promote the expansion

---

OPPOSITE:

A ball rolling down a hill—a mathematical analogue for the dynamics of the inflationary universe. The state of the ball represents the state of the universe. Its height is the energy density in the universe; its position is the magnitude of the spinless field that drives the inflation; its energy of motion is proportional to the sum of the pressure and the energy density in the universe. The shape of the hill depends on the temperature of the universe. At a very high temperature, the ball sits in the bottom of a well. As the temperature falls, the well assumes the shape of the hill shown and the ball begins to roll down the hill (the inflationary phase) and then comes to rest at the bottom (the big-bang phase). The bottom illustration indicates what might happen if we are living in

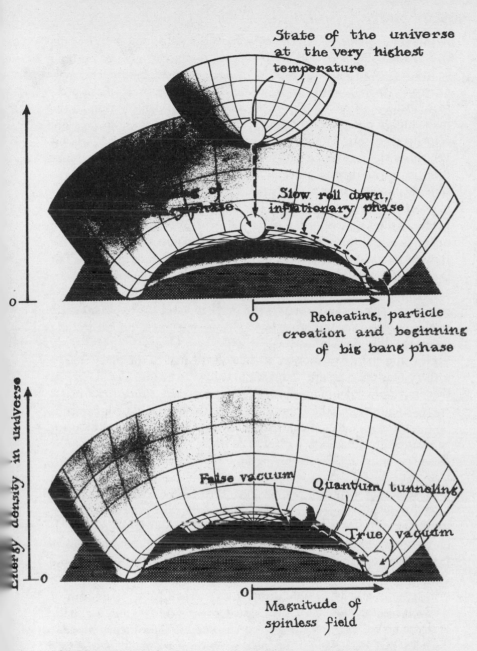

a false-vacuum state. In such a state the true bottom has not yet been reached. By a process of quantum-mechanical tunneling, our universe would then decay into the true vacuum.

in just the way increasing the pressure inside a balloon increases the size of the balloon. But the pressure in the universe is not pushing against any wall—the universe does not have a wall—and the analogy to the balloon pressure is wrong. In fact, one effect of the pressure in the universe is to produce an additional contribution to the mass-energy density in the universe. Hence an increase in pressure actually slows down the expansion rather than speeding it up. Conversely, if the pressure decreases, the expansion rate increases. Since there is no rule against a negative pressure in the universe, it is possible to imagine the pressure's having negative values, and this implies an even faster expansion rate. If the pressure becomes equal to precisely the negative of the mass-energy density (which is assumed not to be zero), then something remarkable happens according to the Einstein equations: the radius of the universe expands at a very fast exponential rate—a truly inflationary expansion.

Now that we understand the qualitative relation between the mass-energy density and pressure on the one hand and the expansion rate of the universe on the other hand, let us return to the ball rolling down the hill.

Here the energy of motion of the ball is equal to the mass-energy density plus the pressure. Thus if the ball were to roll very slowly so that its energy of motion is negligible, then the sum of the mass-energy density plus the pressure is effectively zero. The pressure is then the negative of the energy density—precisely the condition for an exponential inflationary growth of the radius of the universe. The existence of an inflationary expansion of the universe is thus associated with a very "slow rollover" for the motion of the ball. What kind of shape for the hill will produce such a slow rollover?

Before answering this question we have to consider one further complicating feature of the hill that is important: its shape depends on the temperature of the universe and changes as the temperature falls in a way that can be calculated. Before the inflation begins, the universe is extremely hot and the shape of the potential hill is a deep well (as shown). The stable position of the ball is at the bottom of the well. As the universe expands, its temperature falls and the shape of the potential hill, according to certain specific field-theory models, assumes that of a flat central plateau with a gradually descending slope to a "cliff" which falls to zero and then rises again steeply. Consequently, as the temperature falls, the ball, representing the state of matter in the universe, starts out located at the top of a plateau—

not a stable location. Physicists refer to this state of the ball as a "false vacuum state" because it represents a universe which is unstable. The stable position of the ball is at the bottom of the hill—the "true vacuum state."

Random quantum fluctuations in the spinless field will start the ball slowly rolling down the plateau. Because the energy of motion of the ball is negligible during this slow rolling period, the pressure is equal to the negative of the mass-energy density, and this implies that the universe is inflating to an immense size. Hence the existence of the long flat plateau is crucial to making the inflationary idea work. When the ball finally starts rolling quickly down the "cliff," the energy of motion of the ball is no longer negligible and the inflationary phase is over. A new "reheating phase" has begun.

The spinless field that is represented by the rolling ball interacts with the other quantum fields—the gluons, quarks, leptons and Higgs fields—and this interaction can be represented by a contribution to the "frictional" force on the motion of the ball as it falls down the "cliff." As a result of "friction" and reheating, the ball oscillates back and forth in the valley, losing energy, and eventually comes to rest at the bottom of the valley—the true stable vacuum. There are now zero energy density and zero pressure associated with the spinless field; it no longer dominates the matter in the universe. What happened is that the immense energy originally in the spinless field (when the ball was on the plateau) has been used to create the primordial big-bang fireball of all the quantum particles. During the inflation the universe was relatively cold, but the "frictional" force has reheated the now-inflated universe to an immense temperature and the true big-bang phase has begun. The spinless field, having done its work by inflating the universe, is laid to rest, but its offspring is the radiant fireball of quantum particles. The inflation has also created the immense entropy we see today in the microwave radiation—it is the true "heat death" of the universe. After reheating, the universe continues expanding according to the slower rate associated with the standard big-bang model of a radiant gas of quantum particles. The hot big bang has begun.

In summary, the inflationary-universe scenario begins with the universe in an extremely hot state with the matter density dominated by a single spinless field. The temperature rapidly drops, but the universe ends up in a "false vacuum state" with its size inflating exponentially. The inflation stops after the "slow rollover" is completed and the immense energy density of the spinless field drops to zero, finding the "true vacuum." The energy in the field during the reheating phase creates the hot big-bang gas of quantum particles. The universe is now 10^{-33} seconds old.

The major task of the theoretical physicists working on the inflationary-universe idea is to find a relativistic quantum-field-theory model which has a spinless field and for which the calculated shape of the hill is just right so that the scenario works in all its details,

Guth's original version of the inflationary universe was based on the SU(5) GUT, and this model, although it implied inflation, had the unfortunate further property that the inflation did not last long enough to produce the homogeneous universe we see today. Guth was aware of this problem and stated it in his paper. Subsequently Andrei Linde in the Soviet Union, and independently Andreas Albrecht and Paul Steinhardt at the University of Pennsylvania, discovered that other field-theory models had the required hill with a long plateau which provided the slow rollover and the needed prolonged inflation. Since that time physicists have invented many other field-theory models based on GUTs and supersymmetry ideas, all with the intent of improving the inflationary scenario and deriving it from first principles. Many theorists are convinced that the inflationary universe lasting from the first 10^{-35} seconds to about 10^{-33} seconds is logically required by the field theories implementing the "wild ideas." The inflationary-universe idea is the major offspring resulting from the recent marriage of cosmology and quantum-field theory.

Theorists who accept the success of the inflationary-universe idea are applying it to still other cosmological puzzles. For example, inflation and the subsequent reheating phase provide the key to a solution of the matter–antimatter asymmetry puzzle. Recall that one of the requirements that this asymmetry be produced is that the early universe be in a nonequilibrium phase, during which the expansion rate of the universe was fast compared with the collision rate between quantum particles. The inflationary phase satisfies this crucial requirement. Theorists estimating the matter–antimatter asymmetry produced during the inflationary and reheating phase found roughly the observed asymmetry. Without such a nonequilibrium phase (which occurs naturally in the inflationary scenario but not in the subsequent big-bang phase), it is difficult to understand the observed origin of the matter–antimatter asymmetry.

Physicists are also excited because the inflationary-universe scenario may provide the solution to the riddle of the origin of the galaxies. Even before the inflationary scenario was discovered, they knew that galaxies could have grown from tiny fluctuations in the density of matter in the very early universe. But the origin of such fluctuations was always a puzzle.

One solution proposed by many physicists is that these fluctuations were quantum fluctuations in the gas of quantum particles. What is attractive about this solution is that such quantum fluctuations are always present—there is no need to explain where they come from. Unfortunately, detailed calculations showed that these fluctuations were too small by factors of billions to have gone on to produce galaxies in the big-bang model. When the inflationary-universe scenario was discovered, physicists realized there was a new way of examining the problem. Ordinary random fluctuations in the spinless field that produced the inflation could, during the inflationary period, produce the needed density fluctuations to subsequently make galaxies.

In 1982, a number of physicists attending the Nuffield Conference in England calculated these density fluctuations due to the spinless field and all reached the same conclusion: the fluctuations were now too big by factors of 1,000! At the time even that wrong result seemed like progress, since all the previous attempts to calculate the size of the fluctuations resulted in estimates far too small. Subsequently, clever physicists found field-theory models that even gave the right size for fluctuations. For the first time, rather simple theoretical models based on field theory could explain the origin and evolution of the fluctuations from which galaxies eventually formed. Not only does inflation explain the overall uniformity of the universe: it also correctly determines the deviations from that uniformity—the lumpiness of the universe we see today as galaxies. If correct, this implies that the magnificent galaxies began as almost inconspicuous field fluctuations in the first 10^{-33} seconds of our universe.

Besides offering new insights into the problem of the origin of galaxies and the matter–antimatter asymmetry, the inflationary scenario also emphasized a potential disaster that might await our universe. After the inflation, the spinless field comes to rest in the valley at the bottom of the hill—the "true vacuum." But how do we know that this is really the "true vacuum" and that there isn't a deeper valley on the other side of the hill? The answer is we don't know.

It is possible to imagine and build mathematical models which imply that today we are living not in the true vacuum but only in yet another false vacuum. If that is so, then according to quantum theory there is a small probability that after many years our false-vacuum universe will decay into the lower true vacuum—the ball can tunnel right through the hill that separates the two valleys. Sidney Coleman and Frank de Luccia at Harvard University studied this possibility and concluded that the universe after it "tunneled through" to the true vacuum would live less than a few microseconds before collapsing. Reflecting on this, they remark:

> This is disheartening. The possibility that we are living in a false vacuum has never been a cheering one to contemplate. Vacuum decay is the ultimate ecological catastrophe; in a new vacuum there are new constants of nature; after vacuum decay, not only is life as we know it impossible, so is chemistry as we know it. However, one could always draw stoic comfort from the possibility that perhaps in the course of time the new vacuum would sustain, if not life as we know it, at least some structures capable of knowing joy. This possibility has now been eliminated.

Cosmology, not economics, is surely the "dismal science."

The pre–big-bang inflation of the universe and the subsequent reheating phase solve many long-standing cosmological puzzles—the origin of the isotropy and flatness of the universe, the absence of magnetic monopoles and possibly the matter–antimatter asymmetry and the origin of galaxies. Wild ideas like the GUTs that first promoted the inflationary-universe scenario may not survive the test of time. But the central idea of inflation—that the universe at one time underwent an immense expansion—may survive in view of the current failure to find any alternative solutions to the outstanding cosmological puzzles posed by the observed universe.

If we assume the validity of the inflationary scenario, can we then imagine what might have taken place *before* inflation—the very origin of the universe? Inflation stretches out the universe to an almost unimaginable degree and dilutes all matter and structure within it. Indeed, one of the beautiful features of this scenario is that inflation can take almost any hypothetical initial conditions for the universe to the right beginning state for the standard big-bang model; "fine tuning" of the initial conditions is not needed. But this very feature means that no structures can survive inflation that might provide us with "clues" about the preinflationary universe the way the uniformity and flatness of space were "clues" about the existence of an inflation.

Only the existence of the whole universe, its three spatial and one time dimensions, can be viewed as "clues" to the origin of the universe. The fact that the universe exists must be accounted for. In the next chapter we examine the modern speculations on the ancient puzzle of how the world began.

5

Before Inflation: The Origin of the Universe

I am encouraged to believe that the origin and properties of our Universe may be explicable within the framework of conventional science.
—Edward P. Tryon, 1973

How did the universe begin? For centuries, reflective people have contemplated this question. Some believe that the answer must lie outside the domain of science; they feel the creation of the universe was a divine act. Others dismiss the question entirely, maintaining that the universe never began and that it has always existed—a view most recently expressed by the steady-state model of the universe. Yet all the astronomical evidence supports the simple fact that our universe was very different in the remote past and had a definite origin. Conceivably, the universe is endlessly periodic, undergoing expansion, contraction and reexpansion. But this periodicity, if true, cannot be established on the basis of present observations. While a recycled universe is a possibility, we are not compelled to adopt it, and for simplicity we assume that the origin of our universe is a unique event.

If we look at the universe according to the usual cosmological models, the temperature and the density of matter continue to rise without limit as we go backward in time. Eventually the space-time singularity is encountered and the laws of physics no longer make sense. This circumstance leads some people to adopt an attitude that I call "singularity mysticism"—the idea that even scientists must give up the attempt to rationally comprehend the origin of the universe. Such individuals remind me of those who, in the first decades of this century, thought that physics had met its nemesis in the attempt to understand the atom. But the atom was subsequently understood by means of the new quantum theory. Likewise, nothing stands in the way of a rational description of the very origin of the universe, and someday this will be achieved. The singularity at the beginning of time should be viewed as a challenging puzzle, not a signal that we must give up.

But there are even reasonable people who object to the idea of understanding the origin of the universe because it is beyond anything we can expect to subject to observation or test. Furthermore, they would argue, if the inflationary scenario is correct, then any preinflationary features of the universe would have been diluted away and no observational clues to its origin would remain.

But such objections are not valid. Certainly the very existence of the entire universe and the big bang is evidence that there was some kind of origin. There are other features of our universe that may provide information about its origin, though we may not at first think of them as clues. For example, the inflationary picture requires that before inflation the universe was immensely hot and very dense —requirements that should logically follow from a theory of the very origin. Yet another example of a clue is the most dramatic feature of the universe to have survived inflation: the three-plus-one dimensionality of space-time. A further feature of the preinflationary universe is that it exhibits a high degree of symmetry, and this also should be explained by any theory of the origin.

As we embark on the attempt to understand the very origin of the universe, it is worth reminding ourselves of "Einstein's postulational method." This method consists of intuitively guessing a physical postulate (which cannot be directly tested), then logically deducing its consequences and subsequently testing these results against experience. If the tests fail, the assumed postulate must also be rejected.

Scientists will certainly have to guess the correct physical model of the origin of the universe and even the physical laws that govern this event. But interestingly, what we already know about the laws of physics and the observed features of the universe severely constrains our guessing game.

Some scientists are excited because the new ideas coming from quantum physics can be used to make definite mathematical models of the very origin of the universe which avoid a true physical singularity. Such models, like the "wild ideas" they are often based on, are without definite experimental support. But at this very early stage of imaginative model building the lack of experimental support does not trouble physicists. For what is remarkable about these models is not so much that they will ultimately be proved right or wrong but that such rational-mathematical models of the very origin of the universe are possible at all. It seems that the universe, in spite of its immensity and unfamiliar origin, because it is a physical entity will be comprehended by reason.

The very origin of the universe takes places before the inflationary period. As the universe continues to contract it gets hotter and denser and, according to classical general relativity, collapses into a space-time singularity. But this purely classical picture of collapse must be modified if quantum theory is taken into account. Physicists know that the classical description of space-time geometry breaks down at the Planck-length scale before the singularity is encountered. The geometry of the universe then becomes like a frothing sea which physicists describe as a "space-time foam" and the influence of quantum gravity becomes dominant. Since space and time are the most basic concepts used in physics—analogous to the use of words in sentences—it is difficult to say what remains of the conventionl laws of physics in this strange state of the universe—it would be analogous to losing the meaning of words. Yet physicists have found that they can use a new language of field configurations to describe the very origin of the universe. How do some physicists think about this event?

First I should make it clear that physicists and cosmologists do not agree upon any "standard" model of the very origin of the universe. All the models we will survey here should be viewed as very preliminary and to be discarded as better models are invented. What physicists ultimately seek is a definite model of this event, like the definite models of the interiors of stars or the later stages of the big bang when helium is made.

One criterion that any such model should satisfy is that it not leave open unanswered questions about a "preorigin" state of the universe. Otherwise it is not truly a theory of the origin of the universe. For example, we might want to believe the universe began as "something"—a primordial seed. But then we are left with the question Where did that seed come from? The alternative idea that the universe began as "nothing," a creation *ex nihilo*, satisfies the criterion that it leave no unanswered questions about a preexistent state of the universe. But what is "nothing"?

The physicists' usual idea of nothing, what they call the "vacuum state," is the state of lowest possible total energy for a physical system. Such a physical system could be the solar system, a galaxy or the entire universe. Flat, empty space fits this description of the physicists' "vacuum state." But put anything into that vacuum state like an electron or a photon, then its total energy is increased, and it is no longer a state of nothing; it is not empty.

This definition of the vacuum state—as reasonable as it seems—depends upon our ability to define precisely what is meant by the "total energy" of a physical system. A way to do this is to note that energy is equivalent to mass and mass is a source of a gravitational field. If we go far away from a physical system and measure the gravitational field it produces, we can determine its total mass and hence its total energy. However, an unambiguous meaning can be assigned to this gravitational field produced by the system only if space is flat at great distances from the system. If space is not flat at great distances but curved, then the curvature can be thought of as representing the presence of an additional gravitational field which cannot be disentangled from that produced by the matter. This means that we cannot determine the gravity field which is due to the matter and therefore the energy of the system.

The "total energy" of a physical system, a meaningful concept for flat space, is a meaningless concept if we allow for arbitrary space curvature. Since the space of the entire universe can curve, the total energy of the universe is thus simply *not* a meaningful concept. This conclusion, that the concepts of total energy and total energy conservation do not apply to the whole universe, is quite startling—but true. It implies that if we are to define nothing—the vacuum state—as it might apply to the whole universe, then we ought to look for features of the vacuum that do not use the concept of "total energy."

One feature of the vacuum state is that it should be electrically neutral; it should have no electrical charge or any of the other kinds of conserved charges considered by quantum-particle physicists. For if the vacuum possessed such an absolutely conserved net total charge, then one could never get rid of all the particles that carried the charge and the vacuum would not be "nothing"—it would be something.

If we define a vacuum as the state for which all physically conserved quantities are zero, then we find, amazingly, that the entire universe could be equivalent to nothing. A first reaction to this suggestion is that it surely must be nonsense because the entire universe is everything; it is not "nothing." Yet if we examine this suggestion closely, we find that indeed the universe could be equivalent to a state of nothing and hence it is possible that our universe originated from the vacuum.

The first person to express this idea was Edward Tryon, a former student of Steven Weinberg's and now a physicist at Hunter College in New York City. Ed and I meet from time to time at scientific meetings or encounter each other on the streets of Manhattan's Upper West Side, where we both live. During one of those encounters many years ago, Ed told me about his idea for the origin of the universe—that it could begin from a vacuum—and this stimulated my own thinking about the problem.

Tryon in his 1973 article "Is the Universe a Vacuum Fluctuation?" points out that the sum of all conserved charges, such as electric charge, for the whole universe is consistent with being zero and therefore the universe can be created out of the vacuum. No law of physics prevents a creation *ex nihilo*.

Tryon also suggests that the universe originates as a quantum fluctuation of the vacuum—a tiny fluctuation that turned into the big bang. Just as quantum particles can be spontaneously created out of a vacuum, so too the universe might have been created out of a vacuum. But though a spontaneous quantum fluctuation in a vacuum may momentarily produce a particle and an antiparticle, they are quickly destroyed; this is because the actual production of real particles, since such particles have positive net energy, entails a violation of energy conservation. Although energy conservation can be violated according to the quantum uncertainty relation, it can be violated for only a brief time. How, then, can the whole universe emerge from a vacuum quantum fluctuation if even two particles cannot do this?

The reason that real particles are not popping into existence out of empty space today is that our space is very flat, and for such a space the law of energy conservation forbids such a process. But in the very early universe space was highly curved, and therefore total energy conservation was meaningless. A quantum vacuum fluctuation creating real particles *can* take place if space is highly curved. Such a fluctuation might "run away" with itself, creating the many quantum particles that we now identify with the big bang.

As novel and intriguing as it seems today, Tryon's article had little impact on cosmological thinking until quite recently, probably because he did not go on to make an explicit mathematical model that realized his speculations. In the early 1970s, Y. Zel'dovitch and A. Starobinski in the Soviet Union, unaware of Tryon's work, suggested that the quantum fluctuations in the geometry of space-time during the Planck era could produce particles and antiparticles. These newly created particles would remove gravitational energy from the fluctuating geometry, eventually smoothing out the fluctuations to produce a hot universe as in the big-bang model.

Beginning in 1978, R. Brout, P. Englert, E. Gunzig and later P. Spindel of the University of Brussels produced a series of papers which described the first model of the universe originating from nothing by a quantum process. For them the vacuum state, the state of nothing from which the universe begins, is flat empty space. They show how, if a quantum fluctuation produces a few particles in such a space, the mutual gravitational interactions among them cause the space to become curved. A cascade of particle production then follows and space becomes curved in the process—an open, expanding universe filled with matter is created out of flat empty space. The authors conclude, "In a word, we show that the laws of quantum mechanics formulated in the general relativistic framework are perfectly consistent with the spontaneous creation of all the matter and radiation in the universe. This creation has at inception some arbitrary space-time origin."

But a puzzle remains: once such a vacuum fluctuation gets started, it can keep itself going; but how did it get started?

Clearly, the origin of the universe was a very violent event. Working at Rockefeller University in 1981, David Atkatz and I became interested in this puzzle. We developed a mathematical model in which the origin of the universe is a quantum-mechanical tunneling event similar to the decay of an atomic nucleus when its particles tunnel right through the nuclear barrier that ordinarily restrains them.

The origin of the universe may be viewed as such an event with a more dramatic consequence. The basic idea is that the beginning state of "nothing" is a false vacuum state—like a ball sitting behind a barrier which has a small but finite probability to tunnel right through the barrier to a lower-energy state. In so doing, the false vacuum decays into the big-bang state of interacting particles.

According to our model, the universe had to be spatially closed—a compact universe. This meant that the initial space of the universe—"nothing"—was a space consisting of a tiny, perfect "sphere" with no matter in it. While the space of a perfect sphere is just as good a state of "nothing" as flat, empty space, we were disappointed that (at least in our model) flat space could not decay into a big-bang universe. We had hoped to find a way to account for the Brout-Englert-Gunzig-Spindel scenario—creation from flat space.

However, a universe that begins as a tiny compact ball suggested something else to us: "What we envision is that the Universe began as [a multidimensional compact Kaluza-Klein space]. A four-dimensional subspace of this [multidimensional space] then tunnels into the fireball configuration, leaving the remainder as the observed internal symmetries." In our view, the universe begins as a multidimensional space with a high degree of symmetry. But the universe with this geometry may be unstable and undergo a decay via the tunneling mechanism. As a consequence, four dimensions become "large" dimensions and rapidly expand in size while the remainder—the small dimensions—stay small and today are represented by the "internal" symmetries of the quantum particles.

We offered no explanation for why four dimensions became large instead of seven or eleven. That explanation lies in the future. But if this idea is correct, then the origin of the universe is the event that establishes the observed number of space-time dimensions. This number, it is important to remind ourselves, is one "clue" about the origin of the universe that survives the subsequent inflationary period—the dimensionality of space-time does not get diluted.

J. Richard Gott III, an astrophysicist at Princeton University, devised yet another model, this time for creating open universes. Gott assumes the existence of an exponentially expanding space—similar to that of the inflationary period—called a De Sitter space, after the Dutch physicist who first explored its mathematical properties. Steven Hawking had shown in 1974 that quantum particles—the Hawking radiation—are produced at event horizons like the surface of a black hole. In De Sitter space, such event horizons are omnipresent, and so Gott reasoned that particles must be produced everywhere in such a space, filling it with hot matter. He then showed that within this De Sitter "superspace," bubbles of ordinary open space could form and grow at essentially the speed of light, and these bubbles would fill with radiation through the Hawking process. As a bubble grows it becomes more dilute and its rapid expansion slows down. The world inside such a bubble then looks rather much like our own universe during its big-bang stage. Lots of bubbles could form in the De Sitter "superspace," corresponding to lots of universes, none of which can ever communicate with any other. According to Gott, the observed 3 K microwave background radiation we observe today is the fossil of radiation produced by the Hawking process when our universe was a tiny bubble.

All the models of the origin of the universe I have discussed so far assume the preexistence of some kind of empty space—the vacuum whence it all began. The model of Brout, Englert, Gunzig and Spindel assumed a flat, empty four-dimensional space. Atkatz and I assumed a closed space, perhaps of higher dimension than four. Gott assumed a preexistent, four-dimensional De Sitter space. Alex Vilenkin, a theoretical physicist at Tufts University, was not satisfied with any of these notions of "nothing." "Space is still something," Alex once remarked to me, "and I think the universe should really begin as nothing. No space, no time—nothing." When Alex first mentioned this possibility to me, I said, "What do you mean by nothing?" He just shrugged his shoulders and declared emphatically, "Nothing is nothing!"

In 1983 he wrote a paper, "The Birth of Inflationary Universes," in which he constructed a mathematical model in which "nothing" —no space or time—quantum-mechanically tunneled into a geometry of space and time from which an inflationary stage for the universe could be launched. His work went beyond some similar ideas developed in Moscow in 1982 by P. Grischuk and Y. B. Zel'dovich, who, however, did not propose a definite mechanism. But how can space and time be created out of "nothing"?

I have previously described a visual image for the space of a closed inflationary universe as a perfectly elastic ring—the rubber-band model. Let us use this image for the creation of the universe (here a one-dimensional universe). As time runs backward, the ring shrinks until it becomes a point and then disappears. From our viewpoint in three-dimensional space, the ring is embedded in our space and disappears in our space. But from the viewpoint of a creature we might imagine living on the one-dimensional space of the ring, space doesn't disappear in some higher-dimensional space; it just disappears. Likewise the real four-dimensional space-time of our universe can just disappear into absolutely nothing. If it can disappear into nothing it can also be created out of nothing.

One way to visualize Vilenkin's creation of space and time is to imagine a sphere suspended over an infinite plane sheet. The physical space of the universe—the one-dimensional ring—is described by the intersection of the surface of the sphere and the sheet, which is nothing—not even space. Then the sphere "descends" toward the plane and just touches it—the space of the universe is now a point. As the sphere penetrates the plane and passes through it, the point grows into a circle representing our one-dimensional ring universe. Then, as the sphere continues moving through the plane, the ring grows and then recollapses to a point and disappears into nothing. A circular one-dimensional space has come into existence and then gone out of existence. In Vilenkin's model, the creation of space-time is a tunneling process and the universe does not start as a point. Instead it starts as a ring which tunnels from a state of nothing; it pops into existence and then grows.

A defect of this image of a sphere descending through a sheet is that time always exists. But time can be viewed as another dimension, and a similar image can be made for the creation of the temporal dimension out of nothing. Finally, imagine doing away with the sphere and plane and the three-dimensional space in which we imagine embedding them to get the idea of a true creation of space and time out of nothing—absolutely nothing. Vilenkin concludes:

> The advantages of the scenario presented here are of [an] aesthetic nature. It gives a cosmological model which does not have a singularity at the big bang (there are still final singularities) and does not require any initial or boundary conditions. The structure and evolution of the universe(s) are totally determined by the laws of physics.

I am attracted to the notion of combining Vilenkin's idea with the creation scenario envisioned by Atkatz and myself. The space and

---

OPPOSITE:
An image for the creation of space and time—here the space of a "ring universe"—out of nothing. The space-time of the universe is the intersection of the spherical surface with the plane. This begins as "nothing." It becomes a point when the sphere just touches the plane and then finally contracts back to "nothing."

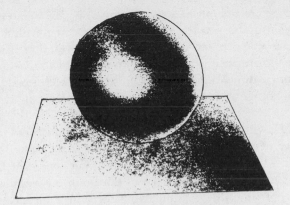

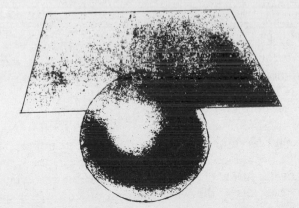

time that are created out of absolutely nothing by Vilenkin's tunneling mechanism are just the multidimensional space-time that is the suggested starting point for us. Once this multidimensional space exists, then, because it is unstable, it tunnels into a three-spatial-plus-one-temporal-dimensional universe. These dimensions are inflated up to an immense size to become the space-time of our physical universe.

Steven Hawking of Cambridge University and James Hartle of the University of California at Santa Barbara carried these ideas a step further. In a 1984 article titled "The Wave Function of the Universe," they directly address the question of the troubling singularity at the beginning of time. Such singularities for which physical quantities like the density of matter become infinite often appear as solutions to the equations of classical physics. For example, the classical equation for a negatively charged electron moving in the electric field of the positively charged proton in a hydrogen atom implies that if the electron falls into the proton, an infinite amount of energy is released. Yet when the hydrogen atom is examined in the light of the quantum theory, no such infinite singularity is encountered. According to the quantum theory, no infinity actually appears because the electron has finite probability of being on top of the proton. The singularity is thus an artifact of a classical description, and if the problem is treated correctly by means of the quantum theory (in which the hydrogen atom is described by a Schrödinger probability-wave function), the singularity is not present.

According to quantum theory, all material things have an associated Schrödinger probability-wave function which describes their state precisely. For most macroscopic objects such as tables and chairs, such a probability-wave description is not very illuminating because large objects can be adequately described by classical physics. For microscopic objects such as atoms or electrons the wave description becomes necessary because the quantum properties of such small things are important. Usually we do not think of the universe as a microscopic object for which quantum properties are important. Yet if we go back in time, the universe contracts until it too became a microscopic object for which quantum effects are important. In order to describe the universe it is necessary to calculate the wave function of the universe just as we ordinarily calculate the wave function of an electron. And just as the wave function of an electron specifies the probability for finding an electron, so too the wave function of the universe specifies the probability for creating a universe.

Hawking and Hartle propose a definition for the wave function of the universe, in particular the specific wave function of the universe when it is in the "ground state"—essentially the vacuum state of the universe. If such a wave-function description for the entire universe makes sense, then they show that the singularity at the beginning of time disappears, just as a wave-function description removes the singularity in the hydrogen atom. They go on to calculate the probability for the universe to emerge from a state of "nothing," as in Vilenkin's model, to the state of "something." Hawking concludes a 1984 follow-up paper, "The Quantum State of the Universe," with the remark "It may well be therefore that the observed universe owes its existence to quantum gravitational effects."

While speculative, like all these ideas about the origin of the universe, the Hawking-Hartle analysis suggests that the origin of the universe can be treated like other quantum events. Theirs is an unfamiliar application of quantum theory because it involves the creation of space and time rather than the creation of a few quantum particles as in the decay of a nucleus. But there is in principle no reason that the origin of the universe cannot be subjected to rational analysis and computation in the quantum theory.

The nothingness "before" the creation of the universe is the most complete void that we can imagine—no space, time or matter existed. It is a world without place, without duration or eternity, without number—it is what the mathematicians call "the empty set." Yet this unthinkable void converts itself into the plenum of existence—a necessary consequence of physical laws. Where are these laws written into that void? What "tells" the void that it is pregnant with a possible universe? It would seem that even the void is subject to law, a logic that exists prior to space and time.

It is easy to criticize each of these ideas about the origin of the universe on technical grounds. But some people even object to the whole investigation with the argument that since the origin of the universe is an unobservable event, it lies outside the domain of science. Those same people, however, would have objected to the calculations of the relative abundance of the elements in the big bang because the big bang is unobservable. Such objections have little merit.

More serious critics would point out that all these models involve the stretching of current theories and concepts far beyond where they have been tested and hence are no more than imaginative guesses. These critics could well be right. A whole community of very smart scientists may have talked themselves into a theory of the very early universe that in the future (with the wisdom of hindsight) will be seen as a fantasy based on incomplete information and imaginative extrapolation. Theory building, while it creates a framework for thought, is never a substitute for experiment and observation. The new high-energy accelerators and telescopes currently on the drawing boards will tell us a lot about whether or not these ideas are correct.

Sometimes I wish that this book about the current ideas of physics and cosmology could be published like a loose-leaf notebook. That way, pages could be discarded and replaced with new pages decribing better ideas when they come along. Much of our current scientific thinking about microscopic physics, the "wild ideas" and cosmology is probably wrong and will have to be discarded. Maybe in the future there will be a major revolution in physics that will revise our whole idea of reality. We may look back on our current attempts to understand the origin of the universe as hopelessly inadequate, like the attempts of medieval philosophers trying to understand the solar system before the revelations of Copernicus, Kepler, Galileo and Newton. What we now regard as "the origin of the universe" may be the temporal threshold of worlds beyond our imagining. But it is also possible that we are near the end of our search. No one knows.

People once worshiped the sun, awed by its power and beauty. Now that astrophysicists understand the physics of the sun and stars and the source of their power, they are no longer the mysteries they once were. In our culture we no longer worship the sun and see it as a divine presence as our ancestors did. But many contemporary people still involve their deepest feelings with the universe as a whole and regard its origin as mysterious. The size, splendor and glory of the universe still provoke the sense of transcendent eternal being.

Someday (and that day is not yet here) the physical origin and the dynamics of the entire universe will be as well understood as we now understand the stars. The existence of the universe will hold no more mystery for those who choose to understand it than the existence of the sun. Steven Hawking, who has contributed so much to the modern understanding of the universe, startled many of his colleagues in his inaugural lecture of the Plumean Professorship at Cambridge entitled "Is the End in Sight for Theoretical Physics?" He said that the major problems of the universe may be solved in several decades.

Perhaps, if a comparison is to be made that reflects my own optimism, scientists are now at the stage of their understanding of the universe that Eddington was at in understanding the stars when he wrote *The Internal Constitution of Stars* in the 1920s. Eddington, with great conceptual power, applied the then-known physical laws to the understanding of stars and even suggested that nuclear energy (at that time not understood) was the source of their power. He was right in broad outline and even many details, but the modern theory of stars had to await the development of nuclear experimentation and theory. If such a comparison is valid, then we may already understand, in broad outline, the very origin of the universe.

Many reasonable scientists will disagree with this optimistic appraisal. But irrespective of where one stands on the question of how close we are to achieving our goal of understanding the universe, all scientists agree that the important thing is to keep working toward it.

Our growing familiarity with these new ideas should not distract us from realizing how strange they were just a few years ago. When historians of science look back on the 1970s and '80s they will report that for the first time scientists constructed rational mathematical models based on the laws of physics which described the creation of the universe out of nothing. And that will mark the beginning of a new outlook on the creation of existence.

Four

Reflections

In the time when Dendid created all things,
He created the sun,
And the sun is born, and dies, and comes again;
He created the moon,
And the moon is born, and dies, and comes again;
He created the stars,
And the stars are born, and die, and come again;
He created man,
And man is born, and dies, and never comes again.
                                    —African song

1

The Cosmic Computer

There is an infinite number of possible universes, and as only one of them can be actual, there must be sufficient reason for the choice of God, which leads Him to decide upon one rather than another.
—G. W. Leibniz, *The Monadology*, 1714

Even as scientists continue to provide a precise and coherent account of the universe, further interpretive questions about their findings can always be asked. Philosophers, poets and writers have created metaphors that interpret the universe, metaphors that many people, including scientists, have in the back of their minds as they reflect about reality. For example, before the rise of modern science many people viewed the universe as an organism or as a divine revelation. After Copernicus, Kepler, Galileo and Newton the image for the universe shifted dramatically—it was a great clockwork, initially set in motion by a divine hand, but now running on its own and completely determined into the future. With the discovery of quantum theory and the statistical nature of atomic events, this image of a deterministic clock became inadequate to describe the universe. These discoveries imply the need for new metaphors.

An important feature of the quantum theory not shared by the earlier classical theory is that the information we obtain about the world by measurements depends on how we decide to obtain that information—the method of measurement. Quantum theory emphasizes information, its representation, and its transformation. Since computers also transform information, an interesting image for the quantum universe is that it is a giant computer—an information-processing system.

In this metaphor of the universe as a cosmic computer the material things in the universe, the quantum particles, are the "hardware." The logical rules these particles obey, the laws of nature, are the "software." The universe as it evolves can be viewed as executing a "program" specified by the laws of nature although it is not a deterministic program like those in digital computers. What the ultimate "output" of this cosmic computer will be remains to be determined. But we already know that its program has given rise to complex "subroutines" that we can identify with life. So complicated are these subroutines that they seem to take on a life of their own, independent of the cosmic computer.

It is fun to explore the metaphor of the universe as a computer and examine what the relation of this cosmic computer is to that other computer, the soft machine inside our head—the brain. The brain is certainly not a digital computer (nor is the universe); but it *is* an organ that transforms information. Our brain-computers are part of the cosmic computer and are also trying to understand its ultimate structure. The difficulty we experience in doing this might be seen as an "interface problem"—matching computers to each other. Remarkably, in spite of this interface problem it seems that we are able to comprehend the universe.

At first that seems paradoxical—how can the part comprehend the whole? But the brain is a far more complex computer than the macroscopic universe. And it is certainly possible for a complex computer to make a mathematical model of its own simplest parts and the rules they obey. There are no self-referential problems or paradoxes until a computer tries to examine its own internal programs—and that is not what we are doing when we examine the universe's hardware.

One reason that we may succeed in comprehending the universe completely—know its basic hardware and software—is that the universe may be macroscopically and microscopically finite. A recurrent theme in the history of physics is that whenever infinities show up in our mathematical descriptions of physical quantities, it is a sure sign that we do not yet understand what is going on. When a better understanding is reached, the infinities are not there. Nature avoids infinities.

An example is Max Planck's discovery of the quantization of energy exchange. The classical law of the distribution of the wavelengths of light emitted by a red-hot coal implied that the total radiation energy is infinite—an absurd conclusion. Planck modified the classical law by assuming that radiation-energy exchange was quantized—it came in little bundles. This removed the absurd infinity, matched the observed distribution of wavelengths and opened the door to the subsequent quantum theory.

Not long ago, many people believed in the steady-state model of the universe which required that the universe was infinite, possibly in space and certainly in time. But this model is not supported by astronomical evidence, and instead the evidence favors the big-bang model for which even time is finite and had a "beginning." Our universe may be finite; at least, there is no major evidence to the contrary.

If the universe, our cosmic computer, is finite although extremely large, then our cosmology begins to bear a resemblance to the circumscribed world view of medieval times. Ironically even in the face of enormous scientific discoveries, what we may be learning about is the ultimately finite and limiting conditions of existence in our universe. However, this time it will not be a provincial idea of God and the universe that places the restrictions on what is possible; it will be the laws of nature themselves.

Should scientists someday soon understand the basic laws of the creation and subsequent evolution of the universe, then, as Steven Hawking emphasized in his Plumean lecture, the end is in sight for theoretical physics. Many physicists disagree with this opinion and point out its similarity to that held by eminent classical scientists at the end of the nineteenth century who also thought physics was coming to an end. These physicists think that physics will go on forever and that there are endless structures and new laws which remain to be discovered. Who is right?

No one can say for certain. But it is certain that the answer to this question will not come from theoretical speculations, no matter how compelling they seem. The only touchstone for empirical truth is experiment and observation. A few experiments or new observations could destroy the picture of the universe developed up to now—or they could add further support.

Suppose, for the sake of argument, that scientists find the master law of the universe, the basic software of the cosmic computer. Then although physics as we know it will come to an end, there will remain at least two frontiers of research. One is "the frontier of complexity"—the boundary of knowledge about the complex ways in which matter organizes itself into living and nonliving forms. It is one thing to know the basic laws of physics and quite another to deduce their complex consequences.

Another frontier might be called "the frontier of simplicity." The master law of physics will probably be rather simple (otherwise we cannot find it). It will be difficult to resist the intellectual impulse to try to understand why that particular law applies and not some other, equally simple law. Here the metaphor of the cosmic computer may be of help. By viewing the laws of the universe as software I foresee the possible future merger of physics with information science, a branch of mathematics. Information science may be able to tell us if our brains and the cosmic computer are matched in such a way that only one law is consistent with our capacity to comprehend the apparent order of the universe.

There are puzzles raised by the image of the universe as a computer. Usually the laws of nature, such as Newton's law of gravitation, are represented in the universe's hardware, such as the solar system. The laws of quantum physics are even represented in "empty" flat space by quantum particles and antiparticles fluctuating into and out of existence. But suppose the universe begins as nothing —that utter emptiness I previously set forth for which not even space and time exist. In the complete void there is no hardware. Without hardware there can be no software. How are the laws of the universe represented in that void? What "tells" the void to execute the program to convert itself into the plenum of existence?

According to the current mathematical models of the origin of the universe out of nothing, there *is* something present in that void—a probability described by "the wave function of the universe." Quantum theory requires that for every physical entity, be it an atom or a bullet, there is an associated "wave function" which specifies the probability for the results of measurements carried out on that entity. An important feature of this wave-function description is that we must conceptually (as well as in practice) separate the physical entity —the "observed"—from the measuring apparatus—the "observer."

The wave function is not a material thing at all; it simply specifies probabilities for material events. An analogy can be made to rolling dice—a material event—whose results can be described by a probability distribution. The probability distribution is like an "invisible hand" that seems to influence material events.

At first it seems unnecessary to contemplate "the wave function of the universe" because the universe is so large and a classical description with certainties instead of probabilities is valid. However, if we go back in time to the big bang the universe contracts down to the size of a quantum particle, and then the wave-function description is important. But where is the separation between the observer and the observed? The universe includes everything, even the observer; there can be no such distinction.

This problem suggests to some physicists that the usual idea of a wave function simply breaks down for the whole universe. The physicist John Wheeler holds the view that the evolution of subsequent observers through the act of observing the universe creates a texture of meaning which becomes the universe—a scenario he calls the "participatory universe." Other physicists, like Steven Hawking, James Hartle and Alex Vilenkin, feel that there is no deep problem and go on to invent mathematical prescriptions for calculating the wave function of the universe.

If we accept the idea that the quantum universe is described by such a wave function, then this wave function specifies the probability for creating a universe of space-time and matter out of nothing. In a sense, even absolute nothingness is pregnant with the plenum: there is a probability for nothing to become something. But which "something"—what kind of universe—gets created?

This question about the origin of the universe puzzled Einstein. He asked whether God had had any choice in creating the universe the way He did. Einstein had something definite in mind when he asked this question—whether or not the laws of nature would ultimately leave any fundamental physical constants unspecified. If all the fundamental constants are specified, then God had no choice. However, if some constants are not specified, then other possible universes can be created—the cosmic computer can run other programs.

The possibility of other universes has always intrigued people who have wondered why our particular universe exists. This was clearly on the mind of the philosopher G. W. Leibniz when, in 1714, he wrote in *The Monadology*: "There is an infinite number of possible universes, and as only one of them can be actual, there must be sufficient reason for the choice of God, which leads Him to decide upon one rather than another. And this reason can be found only in the fitness, or the degree of perfection which these worlds possess." The implications of Leibniz' reasoning did not escape the notice of his contemporaries. Voltaire, the philosophical satirist, created in *Candide* the character of Dr. Pangloss, a thinly disguised Leibniz. Dr. Pangloss, ever the optimist, responded to the continual disasters that befell him and his companions by reminding everyone that "this is the best of all possible worlds"—small comfort to the victims.

Leibniz thought our universe is special because it is "perfect." Today, some cosmologists and physicists think our universe is special because it contains life. These scientists reason that other possible universes might be inhospitable to life and hence they would have no philosophers, cosmologists and physicists to even think about why they exist. By contrast, our universe has welcomed beings that can think about the universe into which they were born and wonder about its properties. It seems that the attribute of life selects out of the set of all possible universes a small subset (perhaps only one) that can observe itself.

To develop the argument, these scientists suppose that the other universes can be characterized by different values for fundamental physical constants such as the gravitational constant or the value for the electronic charge. This will influence the physical structures of those imaginary universes, structures that might not permit the evolution of life. Hence, our actual universe may be constrained not to be very different from the way it is because otherwise, we would simply not be here to observe it. In the words of the British cosmologist John D. Barrow, "The observations of cosmological parameters made by astronomers are the victims of an all-embracing selection effect—our own existence." This idea that the existence of life introduces a selective criterion, a "sufficient reason" for physically possible universes, has been given a name—the "anthropic principle."

The anthropic principle seems less like a principle of physics and more like a biological principle resembling Darwin's principle of natural selection, here applied to the whole universe. When as a physics graduate student twenty years ago I first learned about this anthropic reasoning to explain why our universe is the way it is, I became intrigued. Here was a form of reasoning completely foreign to the usual way that theoretical physicists went about their business of searching out the mathematical laws of nature. And yet this selective principle might shed light on those very laws.

As I thought more about the anthropic principle, however, it seemed less like a grand Darwinian selective principle and more like a farfetched explanation for those features of the universe which physicists cannot yet explain. Physicists and cosmologists who appeal to anthropic reasoning seemed to me to be gratuitously abandoning the successful program of conventional physical science of understanding the quantitative properties of our universe on the basis of universal physical laws. Perhaps their exasperation and frustration in attempting to find a complete, quantitative account of the cosmic parameters that characterize our actual universe has gotten the better of them.

Of course, there are eminent, if not reasonable, scientists who do not share my negative opinion about the anthropic principle. We could debate its merits and demerits a long time. But such interminable debate is a symptom of what is wrong with the anthropic principle: unlike the principles of physics, it affords no way to determine whether it is right or wrong; there is no way to test it. Unlike conventional physical principles, the anthropic principle is not subject to experimental falsification—the sure sign that it is not a scientific principle. No empirical resolution of its veracity is possible, and a debate about whether it is true or not could go forever.

Most physicists and astrophysicists pursuing their research simply pay no attention to the anthropic principle. They endeavor to understand the basic properties of matter and the universe in terms of physical principles expressed in the precise language of mathematics, physical principles which lead to testable conclusions. For the most part, their ongoing enterprise has been remarkably successful.

By contrast, the influence of the anthropic principle on the development of contemporary cosmological models has been sterile: it has explained nothing; and it has even had a negative influence, as evidenced by the fact that the value of certain constants, such as the ratio of photons to nuclear particles, for which anthropic reasoning was once invoked as an explanation can now be explained by new physical laws. If we compare the progress made in understanding the fundamental constants through new physical theories with the progress made using the anthropic principle, then we quickly see that no knowledge has been gained by the adoption of anthropic reasoning. I would opt for rejecting the anthropic principle as needless clutter in the conceptual repertoire of science.

My own view is that although we do not yet know the fundamental laws, when and if we find them the possibility of life in a universe governed by those laws will be written into them. The existence of life in the universe is not a selective principle acting upon the laws of nature; rather it is a consequence of them.

Life is part of the program executed by the cosmic computer and, as far as we know, its most complex subroutine. Today information scientists are designing computers that possess subroutines which can alter the program of the computer itself. Is it possible that life, or whatever it may become, can alter the program of the cosmic computer, changing the course of its destiny?

It will take more than a metaphor to answer that important question; it will take a far deeper understanding of life and the cosmos than we currently possess. Yet the desire to know the answer to such questions about our destiny will never go away. And that desire is perhaps the profoundest program in our cosmic computer so far.

2
First-Person Science

I can live with doubt and uncertainty. I think it's much more interesting to live not knowing than to have answers which might be wrong.
—Richard Feynman, 1981

Isaac Newton, the discoverer of classical mechanics, was a Unitarian —a belief he kept secret while living in the trinitarian Christian culture of Cambridge University. Although Newton was deeply involved with religious ideas, he kept them out of his scientific writings. Einstein, who held developed, if unusual, opinions and had his own view of the impersonality of the universe, always separated these outlooks from his scientific research. The segregation of privately held beliefs and opinions from one's scientific work is more than simple professionalism—it lies at the essence of the scientific enterprise.

Individual scientists bring many personal motivations to the conduct of inquiry. They may hold peculiar views of reality and unconventional political philosophies. They may steal ideas and use their scientific talent as a vehicle for enhancing their social power. Yet the *intent* of inquiry in the natural sciences is to discover the world order, the rules the universe obeys, and to transcribe what is discovered in such a way that its truth can be reestablished by all competent individuals irrespective of their culture, politics, race or sex. It is a remarkable property of our universe that this intent can be achieved at all.

The picture of the world that results from such scientific inquiry I like to call "third-person science." This is the science we see published in professional journals and hear reported at conferences and seminars. Of course this science is "culture-bound" (especially in the choice of subjects suitable for investigation) and is subject to the all-too-human failings of the scientists who do it. Yet its intent, however imperfectly realized, is that this knowledge be true for all of us. If science fails in that intent, it fails in any claim, however provisional, it might have to truth.

Third-person science shows us the world's material order. We certainly did not ask for it to be that way and we certainly did not invent it. Nonetheless, it has now been brought to consciousness, and that is simply how it is. Of course, the world picture of science is far from complete—there is a vast unknown frontier that challenges scientists today. Yet, in just a few centuries, scientific discoveries have dramatically altered the view of the universe held by educated people. And because people believe these discoveries to be true, they cannot but respond to them intellectually and emotionally. Scientists are no exception. Einstein, whose work did so much to create our current view of the cosmos, often expressed his view about how one could relate to the universe. Once he wrote:

A human being is part of the whole, called by us "Universe"; a part limited in time and space. He experiences himself, his thoughts and feelings as something separated from the rest—a kind of optical delusion of his consciousness. This delusion is a kind of prison for us, restricting us to our personal desires and to affection for a few persons nearest us. Our task must be to free ourselves from this prison by widening the whole circle of compassion to embrace all living creatures and the whole nature of its beauty. Nobody is able to achieve this completely but the striving for such achievement is, in itself, a part of the liberation and a foundation for inner security.

This is an example of what I would call "first-person science"—the personal thoughts of an individual interpreting and responding to the reality of the world discovered by science. The feature of first-person science is that it reflects the personal sensitivity and awareness of the writer. The intent of these reflections is not that they be true or meaningful for all people, as in third-person science, but only that they be true to one's own experience, an experience one is willing to share. Neither first- nor third-person science can be a substitute for the other—the intent of the inquiry and reflections is different in each case.

Scientists, in their "first-person" writings, are not privileged in any way. The accidents of their personal history influence their experience of reality as they do for other people. Now I would like to examine a few examples of first-person science.

The recently discovered immense age and size of the universe and the view of the ephemeral nature of our individual lives that these insights support have evoked in people a variety of attitudes—those of the rational person, the spiritual person, the pragmatic builder and the playful explorer. Most of us experience reality not as an illusion but as something absolute and concrete. We are passionate about our experience of reality, and most of us project our hopes or fears onto the universe. Yet such projections of our imagination actually prevent us from exploring that reality with the openness essential for discovery.

We may be tempted to take the attitude of *Homo sapiens*—the person who follows reason alone. Those who trace out the inexorable logic of reason conclude that since ultimately all life will be destroyed, they see no way out; hope, they insist, is an illusion. Bertrand Russell, the English logician, philosopher and humanitarian, eloquently expressed this outlook in his 1923 book *A Free Man's Worship*:

> That man is the product of causes which had no prevision of the end they were achieving; that his origin, his growth, his hopes and fears, his loves and his beliefs, are but the outcome of accidental collocations of atoms; that no fire, no heroism, no intensity of thought or feeling, can preserve a life beyond the grave; that all the labors of the ages, all the devotion, all the inspiration, all the noonday brightness of human genius, are destined to extinction in the vast death of the solar system; and the whole temple of Man's achievement must inevitably be buried beneath the debris of a universe in ruins—all these things, if not quite beyond dispute, are yet so nearly certain, that no philosophy which rejects them can hope to stand. Only within the scaffolding of these truths, only on the firm foundation of unyielding despair, can the soul's habitation be safely built.

Many people would agree, arguing that reason leads only to this conclusion. Steven Weinberg, a theoretical physicist and a Nobel laureate, spoke for many scientists when he wrote ". . . the more we know about the universe the more it is evident that it is pointless and meaningless."

On the time scales of our universe all our acts seem empty. It is as if we had never existed. The simple acts of daily life, the heroism of labor, the visions of political order, the incomprehensible suffering of individuals and multitudes, the hope of religion, the insights and explorations of art and science, the madness, the glory and joy of our species are destined to an oblivion so complete that even the act of destruction will not be remembered. This is not the conclusion of a pessimistic religion or the raving of an unhappy philosopher, but the only rational inference that emerges from our scientific view of the cosmos.

But is Russell right? I'm not convinced. In the first place, it is logically possible that our present view of the universe is seriously wrong or extremely limited—this has happened before and it might happen again. All current views of the universe—such as Russell's—are based, after all, on cosmological models which are themselves inventions of human thought and imagination. At present, for example, no current cosmological model views as important the effect on the future of the universe of intelligent life. Perhaps life will influence the cosmos in an as-yet-unforeseen way. Can life exist indefinitely? Or is it subject to total extinction? On the basis of present knowledge no one can answer such questions with certainty.

Yet even if Russell is right—that all life will be ultimately annihilated—must it follow that we can build our philosophy only "on the firm foundation of unyielding despair"? I do not see the logic in this. Is the purpose of life, if it has one, to promote its endless continuity? I doubt it. Russell, it seems to me, has found a way to justify his own despair—and we all seem to need such justification at times—by projecting it upon his anticipation of the ultimate death of the universe.

An occupational hazard of hard-nosed rationalists is that they sometimes become prone to spiritual or eccentric attitudes as they grow older. I have a view that inside every superrationalist there is a spiritualist struggling to get out. Often when the spiritualist comes out of the closet he emerges in a bizarre and unusual form, a consequence of the long confinement.

Take the case of Auguste Comte, the nineteenth-century founder of scientific positivism. Comte, a superrationalist in the French tradition, deeply loved his housekeeper, who was his mistress. When she died, Comte was emotionally destroyed. He created a cult around her and required his students to worship her. Of course, in his view, this worship was different from that of conventional religions, and was perfectly rational. Comte never knew what hit him.

Those whom Russell would dismiss as deluded fools might reply that death—even the death of the universe—should not be viewed as a meaningless emptiness but as a form of transformation. Such individuals believe in a cosmic consciousness or God—the attitude of *Homo spiritualis*. Such a transcendent being cannot be examined by the natural sciences and can be viewed either positively as the ground of existence (often the case in Western theology) or negatively as the great emptiness (typical of many Eastern religions). These spiritually oriented world views are profoundly appealing. Compared with such views, scientific naturalism seems shallow and destitute.

The ubiquitous human need to find some ultimate foundation to existence seems to me to be organic or biological in origin—like the sexual impulse. If this is so, then denying such spiritual impulses can only result in distortions of our humanity. The forms such spiritual impulses take in different persons—like the sexual impulse—can be shallow or profound, bizarre or beautiful, or almost nonexistent.

One such outlook is expressed by the Belgian paleontologist and Jesuit philosopher Pierre Teilhard de Chardin. He saw the essence of the universe—as Plato did—as a living being in the process of transformation. In his mystical visions of the future he saw all life and matter intertwined in what he called the noosphere, converging to an Omega point, the final realization of all existence:

> The universe is a collector and conservator, not of mechanical energy, as we supposed, but of persons. All around us, one by one, like a continual exhalation, "souls" break away, carrying upwards their incommunicable load of consciousness. One by one, yet not in isolation. Since, for each of them, by the very nature of Omega, there can be only one possible point of definitive emersion—that point at which, under the synthesizing action of personalizing union, the noosphere will render collectively its point of convergence—at the end of the world.

In 1924, he concluded his book the *Future of Man* with these words:

Like a vast tide the Being will have dominated the trembling of all beings. The extraordinary adventure of the World will have ended in the bosom of a tranquil ocean, of which, however, each drop will still be conscious of being itself. The dream of every mystic will have found its full and proper fulfillment.

Are such mystical visions of the future right or wrong? Is that important? They serve the purpose of sustaining hope—enabling those who hold them to go on and work for the survival of the species. Teilhard's words may mitigate the sense of our own suffering by suggesting that it has a deeper meaning than we currently realize. As Simone Weil, the French writer, remarked in her *Notebooks*, "One should identify oneself with the universe itself. Everything that is less than the universe is subjected to suffering."

Others don't engage in lofty thoughts about the end and purpose of the universe at all. The need to create the conditions for human survival—a job, a family, a community, a nation—draws them into a life of practical activity. Their emphasis on the immediate problems of life reflects the outlook of *Homo faber*—the builder or fabricator. Andrew Carnegie, the great American capitalist and philanthropist who consolidated the U.S. steel industry, expressed this view in *The Gospel of Wealth* in 1900:

> Upon speculations as to the future . . . it seems unwise to dwell. I think we have nothing whatever to do with what may come a thousand or a million years hence, and none of us can know what will come. Our duties lie with the present—with our day and generation, and even these are hard enough to discern. The race toils slowly upward step by step; it has even to create each successive step before it can stand upon it, for . . . [if] it attempts to bound over intervening space to any ideal, it will not rise, but fall to lower depths. I cannot, therefore, but regard such speculations a waste of time—of valuable time—which is imperatively required for dealing with the next step possible in the path upward.

The purpose of life as Carnegie saw it is to work to create the conditions for a better life. Many Socialists—despite their opposition to capitalism—would agree with Carnegie's basic message: what matters is improving the life of people living now and the lives of those who will live in the future.

Consistent with its appealing robustness and vitality, the attitude of *Homo faber* disdains intellectual speculation—pragmatism is at the heart of this philosophy. Beliefs, in the view of the practical person, are important only in terms of their function for individual and social behavior. Such pragmatists are unconcerned about the question of ultimate truth, which they insist is provisional.

It is *Homo faber*, not *Homo sapiens* or *Homo spiritualis*, who moves the modern world. Yet some are unable to ignore what the activist disdains—the possibility that all this activity and building will be utterly destroyed. They also embrace the philosophic view expressed by Socrates that "the unexamined life is not worth living."

Still a different attitude toward cosmic reality is that of *Homo ludens* —the one who plays. Often we view the play of children or young animals as a preparation for adult life. But why do so many of us stop playing as we grow older? The child in us, forever exploring, is the source of our creative power. One characteristic I notice in highly creative individuals is their liberation from social and conceptual stereotypes—they create their own exploration of reality, even their own forms of culture. They are radically open-minded, and play and humor, among the highest attributes of cultural life, are the key to this openness.

The activity of *Homo ludens* ought not to be viewed as only the play of children but should also include the play of adults, which sometimes seems like work. Adults can play the game of life quite seriously and yet still see it as a game. Sports and theater are activities of *Homo ludens* which become so involving that we lose sight of the fact that they are games and play. To achieve a sense of freedom in play is not easy, however; the seeming effortlessness is a consequence of training and discipline.

Richard Feynman, a theoretical physicist at Cal Tech, typifies many of the characteristics of *Homo ludens* with his trickster ways, continual joking and unexpected responses. He works hard on physics problems because of the delight it brings him. Never using a concept in physics until he makes it his own, he offers the rest of us inventive new insights. In a 1981 interview, Feynman recalled how he returned to physics research after working at Los Alamos on the atomic-bomb project during World War II:

One day I thought to myself: I haven't done anything important, and I'm never going to do anything important. But I used to enjoy physics and mathematical things. It was never very important, but I used to do things for the fun of it. So I decided: I'm going to do things only for the fun of it.

That afternoon, while I was eating lunch, some kid threw a plate in the cafeteria. There was a blue medallion on the plate. He threw the plate up, and as it came down it spun and wobbled, and I wondered what the relation was between the two. I was just playing; it had no importance at all. So I played around with the equations of motion of rotating things and I found out that if the wobble was small, the blue thing went round twice as fast as the wobble went round. Then I tried to figure out why that was, directly, from Newton's laws instead of through the complicated equations, and I worked that out for the fun of it.

Then I went to Hans Bethe and said: "Hey, I'll show you something amusing." I explained it to him and he said: "Yes, it's very amusing and interesting, but what's the use of it?" I said, "That doesn't make any difference. It hasn't any use. I'm just doing it for the fun of it."

I continued to play with this rotation and it led me to a similar problem of the rotation of the spin of an electron according to Dirac's equation, and that led me back into quantum electrodynamics, which is the problem I'd been working on. I continued playing with it in the relaxed fashion I had originally done, and it was just like taking the cork out of a bottle—everything just poured out. In very short order I worked out the things for which I later won the Nobel prize.

Feynman's playful attitude toward the puzzles of physics implies that he had little interest in mysticism or any fixed view of reality. He goes on to say:

If you expected science to give all the answers to the wonderful questions about what we are, where we are going, what the meaning of the universe is, and so on, then I think you could easily become disillusioned and look for some mystic answer. How a scientist can accept a mystic answer, I don't know. I can't believe the special stories that have been made up about our relationship to the universe at large because they seem to be too simple, too connected, too provincial. People ask me if science is true. I say no, we don't know what's true. We're trying to find out, and everything is possibly wrong.

I can live with doubt and uncertainty. I think it's much more interesting to live *not* knowing than to have answers which might be wrong. I have approximate answers and possible beliefs and different degrees of certainty about different things, but I'm not absolutely sure of anything. There are many things I don't know anything about, such as whether it even means anything to ask why we are here. But I don't have to know an answer. I don't feel frightened by not knowing things, by being lost in a mysterious universe without any purpose, which is the way it really is, so far as I can tell. It doesn't frighten me.

Feynman, perpetually curious, will try almost anything to explore reality. He tells the following story: He was in a sensory-deprivation tank and had an exosomatic experience—he felt that he came "out of his body" and saw the body lying before him. To test the reality of his experience he tried moving his arm, and indeed he saw his arm on his body move. As he described this, he said he then became concerned that he might remain out of his body and decided to return to it. After he concluded his story, I asked him what he made of his unusual experience. Feynman replied with the observational precision of a true scientist: "I didn't see *no* laws of physics getting violated." Indeed, the reliable accounts of such experiences that I have read, as well as my own experience, confirm his perception: "out-of-the-body" experiences no more violate physical laws than does the experience of dreaming.

Feynman's intuition and openness evoke the admiration of many creative scientists. As the reader has probably guessed, my sympathies lie more with the attitudes of *Homo faber* and *Homo ludens*, if for no other reason than that they have liberated themselves from the obsession with certainty and hope. Yet all the attitudes we have considered so far—those of *Homo sapiens, Homo spiritualis, Homo faber* and *Homo ludens* as well as many more—live inside each of us. They are voices directing the course of our lives, calling us down different paths.

People often ask how they may integrate their views about reality, whether they are mystical or pragmatic, with the findings of science. Most of us have a strong impulse to unify our experience of reality as one sort of thing—a universal truth to the universe—and spend part of our lives looking for it. The trouble with looking for some sort of ultimate truth is that we might think we find it—a form of certainty that closes us to new experiences about reality.

"But," someone might ask, "aren't scientists searching for truth and certainty? What about the discoveries of science—the existence of the genetic molecule DNA, the structure of atoms, the properties of light? Aren't scientists certain about the truth of those discoveries?

And what about the truths of mathematics—are they not absolutely certain?" Indeed, we can have confidence in the provisional truth of scientific knowledge. But we must distinguish carefully between scientific knowledge and those world views of reality which, although consistent with our scientific knowledge, represent extrapolations that go beyond it. There can be no universal certainty about the truth of any such world view, which is neither capable of, nor should have, a rational foundation. Such world views are founded on beliefs which reflect the plurality of our experiences and the variety of our needs. Separating our knowledge from our beliefs, our science from our faith, is an accomplishment that we should not destroy by simply collapsing these two components together for the sake of unification.

Maybe there is some final truth to the universe—I do not know. Yet suspending such beliefs opens us to new ways of exploring. Later we can compare our new knowledge and beliefs with the old ones. Often such comparisons involve contradictions; but these, in turn, generate new creative insights about the order of reality. The capacity to tolerate complexity and welcome contradiction, not the need for simplicity and certainty, is the attribute of an explorer.

Centuries ago, when some people suspended their search for absolute truth and began instead to ask how things worked, modern science was born. Curiously, it was by abandoning the search for absolute truth that science began to make progress, opening the material universe to human exploration. It was only by being provisional and open to change, even radical change, that scientific knowledge began to evolve. And ironically, its vulnerability to change is the source of its strength.

The universe haunts me. This sense of the unfathomable beautiful ocean of existence drew me into science. I am awed by the universe, puzzled by it and sometimes angry at a natural order that brings such pain and suffering. Yet any emotion or feeling I have toward the cosmos seems to be reciprocated by neither benevolence nor hostility but just by silence. The universe appears to be a perfectly neutral screen onto which I can project any passion or attitude, and it supports them all.

And where am I? I am in the present, this imperfect moment, trying to remain vulnerable to its intense specificity. There is no other time for me to be or place to go, no cosmic consciousness nor facile mysticism into which I can retreat. In order to see this moment as the fulcrum of all existence, no detail, no imperfection, no impediment of guilt or resentment can remain unacknowledged. I am the witness of this reality—the crack in the cup, the ache in my limb, the background sound of a voice, the love that surrounds me, these very thoughts and feelings as they fill my awareness. No mind that is curious, no soul that wants to liberate itself from the need for certainty, no heart that is open to new experience can turn from the awesome specificity of the world—the very ground of existence. When a student was asked why he traveled so far to study with a certain rabbi, he responded: "It is not to hear him speak on the wisdom of the Talmud that I go, but to watch him tie his shoelaces." In the love of the mundane, the openness to exploration, the play of imagination, the sublimation of aggression into creative activity, the need to communicate with and love other people lie the source of all great poetry, art and science and my private hope for the liberation of the species.

Bibliography

Out of an abundant literature, here is an incomplete list of books that might be of interest to the general reader who wishes to pursue cosmology and astronomy further. Most, but not all, of these books express the outlook of classical cosmology prior to the exciting developments that now relate cosmology and quantum-field theory. However, magazine articles have appeared that do describe research on the early universe, and I especially recommend those appearing in *Scientific American, American Scientist* and *Sky and Telescope* over the last ten years as well as reviews and current reports in *Science* and *Nature*.

Abbott, E. A. *Flatland*, Dover Publications, New York, 1952.

Abell, George. *Realm of the Universe*, Third Edition, Saunders College Publishing, Philadelphia, 1984.

Abetti, Giorgio. *The History of Astronomy*, Abeland-Schuman, London and New York, 1952.

Allen, David A. *Infrared, the New Astronomy*, Keith Reid, Ltd., London, 1975.

Alpher, R.A., and Herman, R. "Reflections of 'big bang' cosmology," in *Cosmology, Fusion and Other Matters: George Gamow Memorial Volume*, ed. F. Reines. Colorado Associate University Press, Boulder, 1972.

Asimov, Isaac. *Eyes on the Universe*, Houghton Mifflin Co., Boston, 1975.

——. *The Collapsing Universe*, Walker and Co., New York, 1977.

——. *The Universe from Flat Earth to Black Holes and Beyond*, Walker and Co., New York, 1980.

Berendzen, R., Hart, R., and Seeley, D. *Man Discovers the Galaxies*, Science History Publications, New York, 1976.

Bergman, P. G. *The Riddle of Gravitation: From Newton to Einstein to Today's Exciting Theories*, Charles Scribner's Sons, New York, 1968.

Bok, Bart J., and Bok, Priscilla F. *The Milky Way*, Harvard University Press, Cambridge, Mass., 1974.

Bondi, H. *Cosmology*, Cambridge University Press, Cambridge, 1960.

————. *Relativity and Common Sense*, Doubleday, Anchor Books, New York, 1964.

Bova, Ben. *The New Astronomers*, St. Martin's Press, New York, 1972.

Burger, D. *Sphereland: A Fantasy About Curved Spaces and an Expanding Universe*, Thomas Y. Crowell, Apollo, New York, 1969.

Chaison, Eric. *Cosmic Dawn*, Atlantic–Little, Brown, Boston, 1981.

Charon, J. *Cosmology: Theories of the Universe*, McGraw-Hill, New York, 1969.

Davies, P. C. W. *Space and Time in the Modern Universe*, Cambridge University Press, Cambridge, 1977.

————. *The Runaway Universe*, J. M. Dent, London, 1978.

————. *The Forces of Nature*, Cambridge University Press, Cambridge, 1979.

————. *Superforce: The Search for a Grand Unified Theory of Nature*, Simon and Schuster, New York, 1984.

De Vaucouleurs, Gerard. *Discovery of the Universe*, The Macmillan Co., New York, 1957.

Dickson, F. B. *The Bowl of Night: The Physical Universe and Scientific Thought*, Massachusetts Institute of Technology Press, Cambridge, Mass., 1968.

Eddington, A. S. *The Expanding Universe*, Cambridge University Press, Cambridge, 1933. Reprinted University of Michigan Press, Ann Arbor, 1958.

Feinberg, Gerald. *What's the World Made Of? Atoms, Leptons, Quarks and Other Tantalizing Particles*, Doubleday, Anchor Books, New York, 1977.

Ferris, Timothy. *The Red Limit: The Search for the Edge of the Universe*, William Morrow and Co., New York, 1977.

————. *Galaxies*, Sierra Club Books, San Francisco, 1980.

Fritzsch, Harold. *Quarks*, Simon and Schuster, New York, 1983.

Frontiers in Astronomy, Readings from *Scientific American*, intro. by Owen Gingerich, W. H. Freeman and Co., San Francisco, 1970.

Gamow, G. *The Birth and Death of the Sun*, Viking Press, New York, 1952.

————. *The Creation of the Universe*, Viking Press, New York, 1952.

————. *Gravity*, Doubleday, Anchor Books, New York, 1962.

Gardner, M. *The Relativity Explosion*, Random House, New York, 1976.

Geroch, Robert. *General Relativity from A to B*, University of Chicago Press, Chicago, 1978.

Gingerich, O., ed. *Cosmology + 1*, W. H. Freeman, San Francisco, 1977.

Glasby, John S. *Boundaries of the Universe*, Harvard University Press, Cambridge, Mass., 1971.

Greenstein, George. *Frozen Star*, Freundlich Books, New York, 1983.

Gribbin, John. *Our Changing Universe*, E. P. Dutton and Co., New York, 1976.

Harrison, Edward. *Cosmology*, Cambridge University Press, Cambridge, 1981.

Harwit, Martin. *Cosmic Discovery*, Basic Books, New York, 1981.

Hey, J. Stanley. *The Evolution of Radio Astronomy*, Science History Publications, New York, 1973.

Hodge, Paul. *The Revolution of Astronomy*, Holiday House, New York, 1970.

Hoskin, M. A. *William Herschel and the Construction of the Heavens*, Science History Publications, New York, 1963.

Hoyle, Fred. *Frontiers of Astronomy*, Mentor Books, New York, 1955.

———. *The Nature of the Universe*, Mentor Books, New York, 1955.

———. *Galaxies, Nuclei and Quasars*, Harper and Row, New York, 1965.

Hubble, E. *The Realm of the Nebulae*, Yale University Press, New Haven, 1936. Reprint Dover Publications, New York, 1958.

Inglis, Stuart J. *Planets, Stars and Galaxies*, John Wiley and Sons, New York, 1976.

Islam, J. N. *The Ultimate Fate of the Universe*, Cambridge University Press, Cambridge, 1983.

Jammer, M. *Concepts of Space: The History of the Theories of Space in Physics*, Harvard University Press, Cambridge, Mass., 1954. Reprint Harper and Row, New York, 1960.

Jastrow, Robert. *The Exploration of Space*, Macmillan and Co., New York, 1960.

———. *Until the Sun Dies*, W. W. Norton and Co., New York, 1977.

———. *Red Giants and White Dwarfs*, W. W. Norton and Co., New York and London, 1979.

Jeans, J. H. *Astronomy and Cosmology*, Cambridge University Press, Cambridge, 1929.

———. *The Mysterious Universe*, Cambridge University Press, Cambridge, 1930. Reprint Macmillan, London, 1937.

Kaufmann, William J. *Black Holes and Warped Spacetime*, W. H. Freeman, San Francisco, 1979.

Kilmister, Clive. *The Nature of the Universe*, Thames and Hudson, London, 1971.

King, H. C. *Exploration of the Universe*, Signet, New York, 1964.

Kippenhahn, Rudolf. *100 Billion Suns*, Basic Books, New York, 1983.

Laurie, John, ed. *Cosmology Now*, BBC, The Broadwater Press, England, 1973.

Lemaître, G. *The Primeval Atom*, Van Nostrand, New York, 1951.

Menzel, Donald H., Whipple, Fred L., and De Vaucouleurs, Gerard. *Survey of the Universe*, Prentice-Hall, Englewood Cliffs, N. J., 1970.

Mitton, S., *Exploring the Galaxies*, Charles Scribner's Sons, New York, 1976.

Morrison, Philip. *Powers of Ten*, W. H. Freeman, San Francisco, 1983.

Motz, L. *The Universe: Its Beginning and End*, Charles Scribner's Sons, New York, 1975.

North, J. D. *The Measure of the Universe: A History of Modern Cosmology*, Oxford University Press, Clarendon Press, Oxford, 1965.

Page, Thornton, and Page, Lou Williams. *Beyond the Milky Way*, The Macmillan Co., Collier–Macmillan Canada, Ltd., Toronto, 1969.

————, ed. *Starlight*, The Macmillan Co., New York, 1967.

————, ed. *The Evolution of Stars*, The Macmillan Co., New York, 1968.

Pagels, Heinz. *The Cosmic Code*, Bantam Books, New York, 1983.

Pais, Abraham. *Subtle is the Lord . . .*, Oxford University Press, New York, 1982.

Report of the Panels, *Astronomy and Astrophysics for the 1980's*, vol. 2, National Academy Press, Washington, D.C., 1983.

Roman, Colin A. *The Astronomers*, Evans Brothers, London, 1964.

Rood, Robert T., and Trefil, James S. *Are We Alone?* Charles Scribner's Sons, New York, 1981.

Rucker, R. B. *Geometry, Relativity and the Fourth Dimension*, Dover Publications, New York, 1977.

Saslaw, W. and Jacobs, K., eds. *The Emerging Universe: Essays on Contemporary Astronomy*, Cambridge University Press, Cambridge, 1971.

Schatzman, E. *The Structure of the Universe*, McGraw-Hill Book Co., New York, 1968.

Schramm, David N., and Wagoner, Robert V. "What Can Deuterium Tell Us?" *Physics Today*, December, 1974.

Schramm, David N., and Turner, Michael S. "Cosmology and Elementary Particle Physics," *Physics Today*, September, 1979.

Sciama, D. W. *The Unity of the Universe*, Faber and Faber, London, 1959.

————. *Modern Cosmology*, Cambridge University Press, Cambridge, 1971.

Shapley, Harlow. *Of Stars and Men: The Human Response to an Expanding Universe*, Beacon Press, Boston, 1958.

————. *View from a Distant Star: Man's Future in the Universe*, Basic Books, New York, 1963.

————. *Through Rugged Ways to the Stars*, Charles Scribner's Sons, New York, 1969.

————. *Galaxies*, Revised by Paul W. Hodge. Harvard University Press, Cambridge, Mass., 1972.

Shipman, Harry L. *Black Holes, Quasars and the Universe*, Houghton Mifflin Co., Boston, 1980.

Shklovski, I. S., and Sagan, C. *Intelligent Life in the Universe*, Holden-Day, New York, 1966.

Silk, Joseph. "The evolution of the universe: a layperson's guide to the big bang." *Griffith Observer*, July 1975.

————. *The Big Bang*, W. H. Freeman, San Francisco, 1979.

———— and Barrow, John D. *The Left Hand of Creation*, Basic Books, New York, 1983.

Struve, O., and Zebergs, V. *Astronomy in the 20th Century*, Macmillan, New York, 1962.

Sullivan, Walter. *We Are Not Alone*, McGraw-Hill Book Co., New York, 1964.

————. *Black Holes*, Warner Books, New York, 1980.

Taylor, E. F., and Wheeler, J. A. *Spacetime Physics*, W. H. Freeman, San Francisco, 1966.

Toulmin, S., and Goodfield, J. *The Fabric of the Heavens: The Development of Astronomy and Dynamics*, Harper and Row, New York, 1961.

Trefil, James. *From Atoms to Quarks: An Introduction to the Strange World of Particle Physics*, Charles Scribner's Sons, New York, 1980.

————. *The Moment of Creation: Big Bang Physics from Before the First Millisecond to the Present Universe*, Charles Scribner's Sons, New York, 1983.

Verschur, Gerrit L. *The Invisible Universe*, The English Universities Press, London, 1974.

Waerden, B. L. van der. *Science Awakening*, Noordhoft, Groningen, Holland, 1954.

Weinberg, Steven. *The First Three Minutes: A Modern View of the Origin of the Universe*, Basic Books, New York, 1977.

Whitney, Charles A. *The Discovery of Our Galaxy*, Alfred A. Knopf, New York, 1971.

Whitrow, G. T. *The Structure and Evolution of the Universe*, Hutchinson, London, 1959.

Wigner, E. F. *Symmetries and Reflections*, Indiana University Press, Bloomington, 1967.

Wood, Harley. *Unveiling the Universe: The Aims and Achievements of Astronomy*, Angus and Robertson, London, 1967.

Wright, H. *Explorer of the Universe: A Biography of George Ellery Hale*, E. P. Dutton, New York, 1966.

————, Warnow, J., and Weiner, C., eds. *The Legacy of George Ellery Hale*, The MIT Press, Cambridge, Mass., 1972.

Index